Springer Series in Surface Sciences

Editors: G. Ertl, R. Gomer and D. L. Mills   Managing Editor: H. K. V. Lotsch

*) Available as a textbook

Eckehard Fromm

Kinetics of Metal-Gas Interactions at Low Temperatures

Hydriding, Oxidation, Poisoning

With 90 Figures

 Springer

Dr. rer. nat. Eckehard Fromm
Max-Planck-Institut für Metallforschung,
Seestraße 92, D-70174 Stuttgart, Germany

Series Editors

Professor Dr. Gerhard Ertl
Fritz-Haber-Institut der Max-Planck-Gesellschaft, Faradayweg 4 - 6,
D-14195 Berlin, Germany

Professor Robert Gomer, Ph. D.
The James Franck Institute, The University of Chicago, 5640 Ellis Avenue,
Chicago, IL 60637, USA

Professor Douglas L. Mills, Ph. D.
Department of Physics, University of California,
Irvine, CA 92717, USA

Managing Editor: Dr.-Ing. Helmut K. V. Lotsch
Springer-Verlag, Tiergartenstraße 17,
D-69121 Heidelberg, Germany

ISSN 0931-5195
ISBN 3-540-63975-6 Springer-Verlag Berlin Heidelberg New York

Library of Congress Cataloging-in-Publication Data applied for
Fromm, Eckehard,
Kinetics of metal-gas interactions at low temperatures ; hydriding, oxidation , poisoning /
Eckehard Fromm.
p.   cm. -- (Springer series in surface sciences ; 36)
ISBN 3-540-63975-6 (hc. : alk. paper)
1. Metallic oxides. 2. Metallic films. 3. Chemistry, Metallurgy. 4. Oxidation 5. Chemical kinetics.
6. Metals at low temperatures. I. Title. II. Series.
QD561.F76   1998

Cover Design: design & production GmbH, Heidelberg
Production: ProduServ GmbH Verlagsservice, Berlin
Typesetting: MEDIO Innovative Medien Service GmbH, Berlin
SPIN: 10658635   54/3020-5 4 3 2 1 0 - Printed on acid -free paper

Preface

Clean metal surfaces react spontaneously with oxygen in the gas atmosphere and form an oxide coating. At temperatures higher than 500 C oxidation proceeds relatively fast and can be measured by traditional experimental methods such as gravimetry, volumetry or the determination of the oxide layer thickness. A large stock of experimental data is available on equilibria and kinetics of the gas interactions of most metals and many alloys at higher temperatures. Theoretical treatments on the mechanisms of oxidation, compound formation, and solution of impurities from the gas phase in solid and liquid metals are also well developed and published in excellent papers and monographs.

Compared with this situation the knowledge on the kinetics of metal-gas interactions at ambient or low temperatures is incomplete and suffers from the lack of reliable quantitative experimental data. Oxidation takes place as well, but the thickness of the oxide scales is in the nanometer range and the amount of gas molecules absorbed is only $10^{-5}\,\mathrm{g/m^2}$. The determination of such thin layers requires highly sensitive techniques and the initial state of the sample surface has to be prepared by cleaning procedures in ultra-high vacuum systems.

The growth of the natural oxide skin or of other compounds of reactive gases with metals is restricted at low temperature to few atomic distances from the surface because of the limited bulk diffusion rates of oxygen, nitrogen, carbon, or other atoms in metals and solid compounds. Only the diffusion of hydrogen atoms proceeds with rates high enough to enable bulk reactions. However, under normal atmospheric or vacuum conditions the oxide scales present on the metal surface reduce the hydrogen absorption rate drastically or prevent the reaction at all. These facts demonstrate that realistic models on the kinetics of low-temperature metal-gas interactions have to incorporate the surface processes correctly into the mechanisms proposed. The structure of such models can become relatively complex, even when idealistic simplifications are used. Unfortunately, most real phenomena cannot be reduced further without violating basic principles of physics and chemistry. Nevertheless, the general feature of the processes can be simulated by consequent application of standard procedures of classical reaction kinetics. With the aid of numerical methods the mathematical treatment of advanced models composed of three to ten partial steps is no longer a big obstacle.

At a first glance it may be in question why quantitative informations on the formation of such thin, hardly detectable oxide skins on metal surfaces should be of practical interest. But a more close look at the problem shows that the natural oxide scale on metal surfaces plays a crucial role in several fields of traditional and modern technology such as wear and friction, hydrogen embrittlement of steels or poisoning of hydrogen storage materials. The production of modern electronic components by thin-film techniques shifts the size of the structures continuously to smaller dimensions in the nanometer range. Consequently, the formation of oxide scales on metal films with a thickness of several nanometers is a problem that has to be controlled carefully in many manufacturing processes.

Despite the difficulties arising with quantitative investigations on the kinetics of low temperature metal-gas interactions new experimental and theoretical results became available in recent years and improved considerably the understanding of the phenomena. The mechanisms of low-temperature oxidation cannot be described by one of the simpler models, where the time law is approximated by the rate equation for one single partial step, for example diffusion in an oxide layer at high temperature. This makes reading of papers on the subject difficult since some background informations from various fields of physical chemistry and solid state physics are required. This knowledge is normally not at actual disposal to the reader active in other fields and, therefore, the message may frequently miss the addressed scientist or engineer who could take advantage of the data presented or of the trends predicted by models. This situation has stimulated the idea to write a book where typical results of experimental investigations and of model calculations are published together with the elements of chemical reaction kinetics and of some other basic phenomena needed for understanding. Whenever possible obvious and uncomplicated equations have been used for the characterization of the physical and chemical facts. Definitions and formulas required for the analysis of the models discussed here or for development of new ones by the reader are shifted to the appendix chapters. The most important parts of the book are the three chapters where the structure and the results of simple and advanced models on hydriding, on low-temperature oxidation and on poisoning of the hydriding reaction are discussed. Each of these chapters begins with the presentation of some typical experimental results which are then simulated by the models discussed in the subsequent sections.

Scientists and engineers are sometimes troubled by puzzling contamination problems that disturb their experiments or a production process. The examples of experimental results and the theoretical analysis of typical phenomena and mechanisms may be able to give hints what had happened and what could be done to avoid unpleasant effects. Another group of readers may use the models for the interpretation of personal experimental data or for the development of new and better ones by modification or extension of the procedures shown. Last not least, this volume can also be considered as a reference book for basic relations on metal-gas interactions and for approximative rate and time laws of simple processes.

Stuttgart, in July 1998   Eckehard Fromm

Acknowledgments

Writing a book needs stimulation from various sides, discussions with many colleagues, and a lot of technical assistance. I am deeply grateful for the substantial support I received in the course of the work. My first thanks go to Professor G. Ertl and to Dr. H. Lotsch for accepting the book into the Springer Series in Surface Sciences and for the smooth and generous cooperation. Furthermore, I am indebted to the three colleagues who have supplied sections on recent developments in modern experimental methods: Professor J. A. Woollam from the University of Nebraska in Lincoln has written Sect. 3.4 on ellipsometry, Professor H. D. Carstanjen from the Max Planck Institute for Metal Research in Stuttgart Sect. 3.5 on energetic ion scattering, and Dr. U. Klemrad from the Ludwig-Maximilians-University in Munich Sect. 3.6 on X-ray reflectivity. I highly appreciate the improvement of the overview given in Chap. 3 owing to the contribu-tions of these competent authors. A great deal of the material discussed in the book relies on investigations by my former coworkers. I am unable to express personal thanks to all of them; the list would become too long. Nevertheless, I would like to mention at least the names of those who contributed much to the experimental results and to the computer simulations presented in the Chaps. 4, 5, and 6: H. Uchida, H. G. Wulz, H. H. Uchida , H. Naoe, and F. Schweppe were active in the field of hydrogen reactions. H. Cichy, V. Grajewski, and M. Martin have worked on oxidation processes. Finally, I wish to express my thanks to J. Schubert who prepared the figures, and to the staff members of Springer-Verlag and ProduServ GmbH who have been involved in the careful production of the book.

Eckehard Fromm

Contents

1 Introduction

Solid and liquid metals in contact with the reactive gases of the earth atmosphere are examples for unstable chemical systems. The equilibrium state is attained only after the whole volume of metal has been transformed to an oxide. Fortunately, kinetics retards this process in everyday live to rates low enough to avoid substantial problems in the use of metallic objects. Depending on pressure and temperature the elements hydrogen, oxygen, carbon, nitrogen, sulfur, and others can either be dissolved in the bulk of metal phases or compounds are formed on the surface, for example, oxides. The processes of gas absorption as well as the reverse ones of gas desorption or compound decomposition are complex heterogeneous reactions composed of transport processes in the gas phase and in condensed phases and of surface and interface partial reactions.

At high temperature the use of metals is restricted by oxidation or hot-gas corrosion and two types of reaction kinetics are observed: The reaction product can be an adherent and dense solid or a liquid surface layer. Then the problem is diffusion controlled and processes on surfaces and at interfaces are treated as fast partial reactions in thermal equilibrium. If the surface scale is cracking or spalling, the reaction kinetics is described by a heterogeneous metal-gas reaction at uncovered surface areas of the metal sample. For both reaction types and the more complex phenomena of internal or selective oxidation of alloys, a large stock of quantitative data is available in the literature and the theoretical understanding of the reaction mechanisms is well developed and published in many survey papers and monographs based on the classical theories of Carl Wagner.

Another field of practical application is the control of the content of addition elements dissolved in metals by metallurgical and vacuum-metallurgical treatments, for example, oxygen, hydrogen, nitrogen, carbon or sulfur. Also on this subject, data for most metals and many alloys are found in the literature and well-established models exist on the mechanisms of gas absorption and desorption reactions. Since technological processes are normally done at high temperatures, the reaction rate for pieces with usual dimensions is controlled by diffusion, too.

At ambient and low temperatures the situation is quite different. Bulk diffusion is becoming extremely slow for all atomic species if hydrogen atoms are disregarded for the moment. Then metal-gas interactions should be restricted to processes at the surface, that means to physisorption, chemisorption, and heterogeneous catalysis. The reaction products formed on a solid surface should remain there or be released again to the adjacent gas or liquid phase but bulk processes should not directly be involved. In the relatively young but fast established field of surface science, this kind of surface processes is investigated with highly sensitive analytical methods. Detailed informations are available on the structure, bonding states and energies of adatoms on many well-defined surfaces of metals and other materials and the phenomena observed are interpreted by models using the full power of theoretical methods.

However, if clean metal surfaces are exposed to oxygen at ambient or low temperatures, an effect is observed which does not fit this simple classification of metal-gas interactions. Despite the low diffusion rates measured for oxygen atoms in solid state an oxide scale is formed instantaneously with a thickness of several monolayer equivalents by a thermally not activated process. Nowadays this peculiarity of low-temperature oxidation is well known by experimenters since sputter depth profiling is commonly used for the characterization of sample surfaces. An explanation for this phenomenon has been given by N.F. Cabrera and L.N. Mott some forty years ago. The transport of ions in semiconducting surface layers is strongly enhanced by a high electric field which is produced by electrons tunneling from the metal phase to acceptor sites at the oxide surface. This Mott potential enhances the initial reaction rate on clean metal surfaces by orders of magnitude until, beyond a critical thickness of the layer, the tunnel current breaks down. This ever-present and fast self-healing natural oxide skin on metals prevents oxidation so reliably that under normal conditions one does not need to care much about it. And the scientific community did exactly that when the limited research activities are compared with other fields of material and surface science.

The small number of quantitative data available on low-temperature oxidation kinetics is explained by the lack of suitable methods for experimental investigation. The thickness of the oxide skins is in the range of some ten atomic distances and lies exactly between the sensitivity ranges accessible by commercial experimental equipment. For surface analytical methods the layers are very thick and direct determination of weight gain in the range of $10^{-9}\,g/cm^2$ is not easy. Furthermore, it is a problem to prepare and maintain a gas-free metal surface and to measure small quantities of absorbed gases in the monolayer range at higher pressures. Another problem, in addition to the limited depth resolution of surface analytical methods, is the fact that most experiments published so far have been performed at very low pressures where the kinetics of metal-gas interactions is controlled by the impingement rate of the gas species. In experiments with a quartz crystal microbalance or with volumetric methods at low pressures, metal film samples are used with not well defined roughness and structure of the surface. But they can at least determine directly the absolute amount of absorbed gas molecules.

The only reaction which proceeds with fast bulk diffusion rates at ambient temperatures is hydrogen absorption and desorption. However, in most metal hydrogen systems the dissociation energy of physisorbed hydrogen molecules is higher than the diffusion activation energy. Therefore, dissociative chemisorption of hydrogen atoms on the surface is very frequently rate determining. The reaction of hydrogen with a clean surface of reactive metals is a rather unrealistic situation outside an ultra-high vacuum system. Under normal conditions the surface is covered by the natural oxide skin or poisoned by other reactive elements and the reaction rate is retarded by orders of magnitude. Poisoning affects the kinetics frequently in an unpredictable manner, for example, hydriding of metal ingots for powder production, hydrogen embrittlement of steels or the loading-unloading cycles of metal-hydride storage tanks.

The reader may ask, why is a book on oxidation and hydrogen absorption published in a series on surface science? The kinetics of metal-gas interactions can be subdivided roughly into diffusion controlled and surface controlled processes. Scale formation in high-temperature oxidation and hot corrosion are-diffusion controlled and interaction of gas molecules on a solid surface acting as a catalyst is surface controlled. However, what can be said about low-temperature oxidation? The establishment of the Mott potential is a reaction between electrons in the bulk and acceptor species on the oxide surface. The equilibrium concentration of ions in the oxide layer can be a result of surface reactions and the transport of these ions through the oxide skin is transport in a high electric field. But one aspect all these processes have in common is that they occur on or in an area very close to the surface. Under this point of view it may be justified to attribute this subject to surface science. Another argument for this classification may be the fact that scientists and engineers active in surface science normally suffer more from surface contaminations than their colleagues from the high-temperature oxidation and corrosion community.

The content of this book is focused mainly on the presentation of models on the mechanisms of metal-gas reactions at low temperatures which describe phenomena normally encountered as contamination problems. If models on reaction kinetics are compared with experimental data, the results must be given as equations for rate and time laws or as simulated oxidation curves. Simplificative assumptions also have to be made with the models here. Considered are only the formation of solid solutions and adherent reaction products. Cracking or spalling of the coating are disregarded as well as gas phase diffusion and non-isothermal processes. Also not included is the nucleation of oxide islands and absorption of the first three layers of oxygen needed to define the surface, a bulk plane and the metal/oxide interface of an oxide for ionic transport mechanisms. These processes belong to the very fast initial region of low-temperature oxidation kinetics which cannot be measured at elevated pressures for experimental reasons. Thus, they have no practical significance outside an ultra-high vacuum system.

The models itself are developed by consequent application of the laws and procedures of chemical kinetics. Complex reactions are composed of individual partial steps such as physisorption, chemisorption, formation of charged species on surfaces and interfaces, formation of reaction products at the reaction front, transport of neutral and charged particles through surface coatings with and without assistance of an electric field. The main problem of modeling is the definition of the reaction partners involved that are the initial and the intermediate species and the reaction product, and their correct linking by partial reactions, rate constants, and transport equations. The constants introduced must have a clear physical meaning and values with a realistic magnitude. In oxidation models published so far, surface processes have not yet been incorporated consequently, probably because the solution of the problem was hampered by mathematical difficulties before the computer age. Advanced models have a relatively complex structure since five to ten partial steps are involved, even in idealistic simplification. Unfortunately, the real phenomena cannot be simplified further without violating basic principles of physics and chemistry. The time laws of the simplest models on low-temperature oxidation have to consider three branches, initial, intermediate, and final. Each of them has another rate determining partial step.

The calculation of simulated oxidation curves and studies on the effects of parameter variations are no big problems if computer-aided numerical methods are used. Simple mathematical expressions have been chosen to characterize the function of individual partial steps of the models in order to avoid confusion created by an overload of mathematics. Incorporation of better suited, but more complex, equations is no big problem for a reader interested in the detailed analysis of a specific problem.

Basic phenomena of metal-gas reactions and mathematical standard procedures of reaction kinetics are reviewed in Chap. 2. This may ease reading of the main body of the book. Those who are familiar with this material may pass on. In a short third chapter the power and shortcomings of experimental techniques are characterized to assist appraisals on the reliability of quantitative experimental data given in the literature. Definitions, equations, some data needed, and a more detailed description of individual partial processes are compiled in the appendices at the end of the book. Thus, the content of the three main Chaps. 4–6 can be focused on the presentation of the structure of the reaction models and the conclusions derived from the results without repeated mention of peculiarities of the partial steps. Typical experimental curves are presented in the first section of each of these chapters. They show the phenomena which have to be simulated by the models. They are taken from extended series of measurements performed in our laboratory and, therefore, details of the experimental conditions are known. This has considerable advantages since comparing data on low-temperature reaction kinetics measured by different authors with different techniques in different laboratories is still a problem.

In Chap. 4 the kinetics of the solution of hydrogen in metals and of hydride layer formation on unpoisoned metal surfaces are presented. These mechanisms are less complicated than oxidation processes since charge effects do not need to be considered and equations for the limits with only one rate determining step can be formulated. They are also suited to describe other processes where surface coatings with metallic electric conductivity are formed, for example, nitrides or carbides on transition metals. To my knowledge a synopsis on such equations is missing in recent literature.

In Chap. 5 advanced models on low-temperature oxidation are presented. They simulate the complete oxidation process, starting from an almost uncovered clean metal surface and ending with extremely slow reaction rates after formation of the natural oxide skin. This chapter may be the most interesting part of the book since it demonstrates new aspects of the problem. Several partial reaction steps are involved in the mechanism with equal importance and no single step can be defined as rate determining. A reader accustomed to relatively simple standard models may at first feel somehow confused. However, the large number of examples presented with parameter variations should be able to eliminate this dilemma.

Some problems of poisoning fast bulk reactions by surface contaminations are discussed in Chap. 6 using hydrogen reactions as examples. These processes are combinations of the mechanisms treated in Chaps. 4, 5.

The computer programs of the models on the hydriding of spherical metal particles and on oxidation are written in turbo pascal and are available. They run on personal computers and calculation of an absorption curve takes normally only few seconds or minutes.

The content of this monograph reflects in many sections the work of my former and present co-workers. Their results and ideas improved step by step our understanding of sometimes puzzling phenomena observed with low-temperature metal-gas interactions. We would be happy to notice that the one or the other example discussed here or in one of the original papers could stimulate the solution of problems in material science or improve production methods in modern technology.

2 Principles of Reaction Kinetics

Chemical reaction kinetics is concerned with the determination of rate laws and time laws for a reaction, for example, the oxidation of a metal sheet. Rate laws describe the reaction rate as a function of process parameters:

$$v = \frac{dm}{dt} = v(p_{O_2}, T, l, \ldots).$$ (2.1)

The mass gain per time unit of the sheet can depend on the oxygen pressure, the temperature, the oxide scale thickness l and other parameters. The time law is the integration of the rate law over the reaction time and gives in our example the weight gain as a function of the time and the other parameters mentioned,

$$\Delta m \int_{t_0}^{t} v \, dt = \Delta m(p_{O_2}, T, \ldots t).$$ (2.2)

If we try to understand the reaction mechanism hidden behind the experimental data on an atomistic scale, we have to develop a reaction model from first principles which is able to formulate theoretical rate and time laws and to compare them with the curves measured. Metal-gas interactions are complex heterogeneous chemical reactions composed of several individual partial reactions. These partial steps can procede parallel or one behind the other and they are linked together by the concentrations of intermediate reaction products. As long as only one of these steps is the bottle neck of the whole process and, therefore, rate determining the rate and time law can be determined easily. We need only the equilibrium constants of preceding and succeeding partial steps and the rate law of the rate determining partial reaction. However, in real systems frequently two or more partial steps are rate determining in different phases of a process. The mathematical treatment becomes complex very fast if a larger number of possible rate determining steps has to be considered and the problem must be solved in most cases by numerical methods. On the other hand, each partial step treated as a reaction being not in eqilibrium increases the number of free parameters and simplifies fitting of experimental data.

Models on reaction mechanisms have an unpleasent peculiarity, namely, that different mechanisms frequently yield identical or very similar rate and time laws. Thus, agreement between experimental and calculated curves does not automatically prove that the model proposed is correct. Additional arguments and experiments are required to check effects predicted by the model for parameter variations. Careful interpretation of the physical significance of parameters and reasonable values introduced for fitting experimental data can improve the reliability of the mechanism proposed. Despite these principal shortcomings the laws and procedures provided by classical reaction kinetics are very helpful tools for the development of reaction models which can simulate the general trends observed in experiments. Very often they even can simulate quantitatively the kinetics of a process.

A reaction model with a sound structure has to combine the partial reactions and transport phenomena of a limited number of species in such a way that the law of mass conservation is not violated at any site of the system and for very long reaction times the equilibrium state of the process must be attained. These requirements must hold also if approximations are inserted such as a single rate determining step or the steady state conception for intermediate reaction products.

In Sect. 2.2 it will be demonstrated how individual partial steps are combined to a model describing the mechanism of a simple chemical metal-gas reaction and which conditions must be obeyed if approximations are applied. In the Appendix A standard reactions are compiled for the species appearing in the models of this book. Equations for electronic currents and ionic transport mechanisms are found in the appendices C and D. The general behavior of individual partial steps is discussed in Sect. 2.3 and also the order of magnitude for some basic quantities which are needed for qualitative considerations or estimates.

2.1 Equilibria of Chemical Reactions

Relations and equations derived from equilibrium thermochemistry are used in many areas of reaction kinetics. In most models equilibrium conditions are inserted for the concentration of intermediate species which are produced by fast partial reaction steps under steady state conditions or they give the final state of the system approached after very long reaction times. Furthermore, one has to remember that the theory of rate processes is based mainly on equilibrium considerations [2.1–4]. On the top of an activation barrier between two equilibrium states the components of the reactants, and consequently also of the reaction products, are considered to form a metastable cluster which is called the transition complex. If this complex is treated as an intermediate reaction product the probability for its formation can be represented by a concentration term of a quasi-equilibrium state. The reaction rate is then an expression proportional to the concentration of the transition complex times a frequency factor determined by lattice vibrations in the range of 10^{13}/s.

Chemical reactions are defined by reaction equations which contain the reactants, that means the reacting species, and the reaction products. The basic quantities in thermochemistry are the energy exchanged with the environment during the reaction and the final equilibrium concentrations and pressures correlated by the equilibrium constant [2.5–8]. Most equilibria are described for isothermal isobaric conditions, that means for constant pressure and temperature. The energy quantity exchanged, the reaction enthalpy ΔH, consists of the change of internal energy of the system, ΔU, and the volume work $p\Delta V$,

$$\Delta H = \Delta U + p\Delta V. \tag{2.3}$$

The equilibrium state in a system is attained when the Gibbs free energy G is a minimum. The change ΔG of a system caused by a reaction must, therefore, become zero

$$\Delta G = \Delta H - T\Delta S = 0, \tag{2.4}$$

where ΔS is the entropy change. The Gibbs free energy G of a system composed of several phases, for instance of the gas phase and one or several condensed phases such as oxides or metallic phases, is the sum of the G^α values of all phases a belonging to the system. Within a phase the G^α value is composed of the partial molal Gibbs free energies or chemical potentials μ_i^α of the components of the phase, which are in most cases chemical elements. The chemical potential of the component i in the phase α, for example, hydrogen dissolved in the Ti-metal phase, is defined as

$$\left.\frac{\delta G^\alpha}{\delta n_i}\right|_{T,p,x_j} = \mu_i^\alpha. \tag{2.5}$$

This is the change of the value of the G function if one mole of the component i is added at constant p, T, and constant compositions x_i of all components j including i.

After introduction of the chemical potentials the calculation of the equilibrium state of a chemical reaction, for example,

$$H_2 + \text{metal M} \leftrightarrows MH_2, \tag{2.6}$$

is very simple. If the sum of the chemical potentials $\mu_{H_2} + \mu_M$ of the consumed reactants is the same as μ_{MH_2} of the hydride formed, the condition $\Delta G = 0$ is fulfilled,

$$\mu_{H_2} + \mu_M = \mu_{MH_2}. \tag{2.7}$$

This is evident since the G value of the gas phase decreases by an amount μ_{H_2} and the metal phase by μ_M whereas the hydride phase is increased by μ_{MH_2} This example demonstrates that the equilibrium condition of any chemical reaction can be obtained simply by replacing in the reaction equation the chemical symbols of the reactants and the reaction products by their chemical potentials.

For thermochemical calculations equations are required for the chemical potentials of the reacting species in different phases as a function of temperature, pressure, and concentration. In thermochemistry two ideal phases have been defined, the ideal gas phase and the ideal mixtures. The chemical potential of a species i in an ideal gas phase is given by

$$\mu_i = \mu_i^0(T, p_i = 1) + kT \ln p_i \tag{2.8}$$

with the standard potential μ^0 as the μ value at the temperature T and the partial pressure $p_i = 1$ in the units chosen. The pressure dependence is given by the second term on the right handside.

For ideal mixtures an almost identical expression holds

$$\mu_i = \mu_i^0(T, x_i = 1) + kT \ln x_i \tag{2.9}$$

with μ_i^0 standard potential of the pure component i. The quantity x_i is the mole fraction $n/\Sigma_j n_j$ of the mixture with n_i number of moles in the phase considered. The second term is the concentration dependent entropy of mixing of the chemical potential.

Real mixtures are presented in a similar form,

$$\mu_i = \mu_i^0(T, a_i = 1) + kT \ln a_i. \tag{2.10}$$

The activity a_i describes the concentration dependence of μ_i. Normally deviations from the equation for the ideal mixture is described by the activity coefficient γ_i,

$$\gamma_i = \gamma_i(T, x_i, x_j,). \tag{2.11}$$

The function γ_i is a parameter which depends on the physics and chemistry of processes which produce deviations of a real phase from the ideal behavior. The introduction of the activity as a new variable into the thermochemical formalism has big advantages. Results obtained for ideal systems can be used for real systems just by replacing the molefractions x_i by the activities a_i in the equations. The functions $a_i(T, x \ldots)$ for the specific system considered can be inserted at the very end of a treatment in the final result if this is of interest.

Equilibrium conditions for a reaction are defined by the mass action law which can be written as

$$kT \sum_i \ln a_i = - \sum_i \mu_i^0 = K(T). \tag{2.12}$$

For the hydride formation reaction mentioned above this reads as

$$\frac{a_{MH_2}}{a_{H_2} \cdot a_M} = \exp\left(\frac{-\left(\mu_{MH_2}^0 - \mu_{H_2}^0 - \mu_M^0\right)}{kT}\right) = \exp\left(\frac{-\Delta G^0}{kT}\right). \tag{2.13}$$

The term ΔG^0 is called standard reaction Gibbs free energy and compiled in tables of thermochemical data for numerous chemical reactions investigated so far [2.5–9]. The equilibrium constant of the mass action law is defined as

$$K(T) = \exp\left(-\frac{\Delta G^0(T)}{kT}\right). \tag{2.14}$$

In most systems discussed in this book, another limit of the chemical potential is of special interest. This is the *ideal dilute solution* of a species. If the concentration of a component in a mixture is very small, about 1% or less, then the activity coefficient γ_i has a constant value depending only on temperature,

$$\mu_i = \mu_i^0(T) + kT \ln \gamma_i(T) + kT \ln x_i. \tag{2.15}$$

It is convenient to add the γ-term to μ_i^0 and write

$$\mu_i = \mu_i^{0\prime}(T) + kT \ln x_i \tag{2.16}$$

with

$$\mu_i^{0\prime} = \mu_i^0(T) + kT \ln \gamma_i(T). \tag{2.17}$$

Table 2.1 Partial molal functions of components in real sytems

function	standard function	mixing function	excess function
$\mu_i = g_i$	$\mu_i^0 = \mu_i^0(T)$	$\mu_i^m = kT \ln x_i \gamma_i$	$\mu_i^e = kT \ln \gamma_i$
h_i	$h_i^0 = -T^2 \frac{\partial \mu_i^0/T}{\partial T}$	$h_i^m = -kT^2 \frac{\partial \ln \gamma_i}{\partial T}$	$h_i^e = -kT^2 \frac{\partial \ln \gamma_i}{\partial T}$
s_i	$s_i^0 = \frac{-\partial \mu_i^0}{\partial p}$	$s_i^m = -k \ln x_i \gamma_i - kT \frac{\partial \ln \gamma_i}{\partial T}$	$s_i^e = -k \ln \gamma_i - kT \frac{\partial \ln \gamma_i}{\partial T}$
v_i	$v_i^0 = \frac{\partial \mu_i^0}{\partial p}$	$v_i^m = kT \frac{\partial \ln \gamma_i}{\partial p}$	$v_i^e = kT \frac{\partial \ln \gamma_i}{\partial p}$

This expression for the chemical potential of an ideal diluted species in a phase is a very convenient standard state for modeling alloying elements dissolved in condensed phases or lattice defects in stoichometric compounds.

The meaning of the chemical potential and the activity of a component i present in a phase α can be illustrated by the following definition: If one mole of the component i is transferred from its standard state to a phase α, which is so big that x_i is not changed remarkably, then the μ value of this mole of the component i is changed by the amount $\mu_i^\alpha - \mu_i^0 = kT \ln a_i$.

If a charged species i is transferred to a phase which has an electrical potential different from the potential of the standard state then a term $z_i e \varphi$ has to be added to the chemical potential. The new quantity for charged particles i is called electrochemical potential,

$$\eta_i = \mu_i^0(T, a_i = 1) + kT \ln a_i + ze\varphi. \tag{2.18}$$

The definitions for some chemical potentials and other partial molal functions used for real systems in modeling are compiled in Table 2.1

2.2 Structure of Reaction Models

Reaction kinetics is a well-established field of physical chemistry [2.1–4]. It provides standard procedures which are reliable and proven tools for modeling and many classical theories are available as elements for the development of the more complex structures of advanced models. The typical features of a model for the simulation of a chemical reaction and the mathematical procedures required for the formulation of rate and time laws can be demonstrated best by the analysis of a very simple mechanism. The formation of a hydride will be discussed as a typical example for a metal-gas reaction.

In a metal-hydrogen system M–H only hydrogen atoms are mobile at low temperature in condensed phases and ionic fluxes can be disregarded. Furthermore, in our simple model it is assumed that adherent hydride layers are formed on a plane sheet of metal and hydrogen shall not be dissolved in the metal phase. Thus, size effects caused by the sample geometry and interactions between competing reactions, for example, hydrogen solution in the metal matrix and hydride formation, can be neglected. Since metal atoms are assumed to be immobile and electric fields cannot exist in a metallic hydride phase, only the transport of hydrogen atoms must be considered. However, even for this simple system the determination of exact rate and time laws entails considerable mathematical problems. They can be reduced remarkably by introduction of a few helpful additional approximations. But also they are unable to yield a general solution which presents the laws as a mathematical equation. Simple expressions for rate and time laws are obtained only for some of the limiting cases. More general solutions of the problem can be obtained only if numerical methods are applied.

2.2.1 Reaction Partial Steps

The chemical reaction

$$H_2(gas) + metal\ M \leftrightarrows MH_2 \tag{2.19}$$

has to be subdivided into five partial steps in order to accomplish a realistic model for the reaction mechanism on an atomistic scale:
1) $H_2(gas) \leftrightarrows H_2$(physisorbed on MH_2),   (2.20)
2) $H_2(phys) \leftrightarrows 2\ H$(chemisorbed on MH_2),
3) $H(chem) \leftrightarrows H$(interstitial in MH_2),
4) diffusion of H in MH_2,
5) $2\ H(in\ MH_2) + M \leftrightarrows MH_2$(at the MH_2/M interface).

With the reaction equations (2.20) a special mechanism has been proposed or, in other words, a model for the simulation of the kinetics of the hydride formation has been chosen. In addition to the two reactants $H_2(gas)$ and M and the reaction product MH_2 two surface species have been introduced as intermediate reaction products, namely, $H_2(phys)$ or adsorbed H_2 molcules and $H(chem)$ or hydrogen atoms adsorbed on the hydride surface after dissociation of the H_2 molecule. Furthermore, hydrogen interstitials in the hydride phase MH_2 have been postulated as lattice defects which accomplish the transport of hydrogen atoms through the hydride layer from the surface to the reaction front at the metal/hydride interface. There, finally new hydride molecules are formed. This mechanism is the most simple reaction model for the formation of an adherent dense surface compound layer by a metal-gas reaction (see also Fig. 4.3).

   The formulation of chemical partial reactions between atomic, ionic and electronic species and of the flux equations for migrating particles and the interrelations between these processes has to be accomplished in such a way that no hidden sinks or sources for any of the species defined are incorporated into the structure of a model. At this early stage of model development, the reliability of the assumptions made should be checked very carefully with respect to the laws of mass and charge conservation.

2.2.2 Rate Equations of Partial Reactions

The next step in developing the model is the formulation of rate equations for the individual partial reactions. The procedure described in text books [2.1–4, 10] finally yields the following rule: The reaction rate v is proportional to the probability of finding all reactants in a local area small enough to permit interaction times a rate constant $k = k^0 \exp(-A/kT)$ which depends on temperature and is defined by a frequency factor k^0 of about $10^{13}\,\mathrm{s}^{-1}$ and an activation energy A. Thus, for each reaction a rate equation for the forward and the backward reaction can be written according to this rule, for example, for a reaction $A + 2B = AB_2$:

$$A + 2B \rightarrow AB_2, \qquad v_+ = c_A \cdot c_B^2 \cdot k_+, \tag{2.21}$$

$$AB_2 \rightarrow A + 2B, \qquad v_- = c_{AB_2} \cdot k_-. \tag{2.22}$$

Equilibrium between A, B, and AB_2 exists if $v_+ = v_-$ holds or

$$\frac{c_{AB_2}}{c_A \cdot c_B^2} = K(T) = \frac{k_+^0}{k_-^0} \exp\left(\frac{-A_+ + A_-}{kT}\right). \tag{2.23}$$

In solid state reactions frequently a *site* for the reaction product is also required. This site must be treated like a separate reactant. The first equation in (2.20) then becomes :

$$H_2(\text{gas}) + \text{site} \rightarrow H_2(\text{phys}), \qquad v_+ = p_{H_2}(1 - \theta_{H_2})k_+, \tag{2.24}$$

$$H_2(\text{phys}) \rightarrow H_2(\text{gas}) + \text{site}, \qquad v_- = \theta_{H_2(\text{phys})}k_-. \tag{2.25}$$

$\theta_{H_2(\text{phys})}$ is the relative coverage of the hydride surface with H_2 molecules ($0 \le \theta_{H_2} \le 1$). The equilibrium condition $v_+ = v_-$ yields the Langmuir equation of physisorption

$$\frac{\theta_{H_2}}{(1 - \theta_{H_2})} = p_{H_2} K(T). \tag{2.26}$$

The rate and equilibrium equations of the partial steps 1, 2 and 3 of (2.20) are compiled in Table 2.2.

The rate law for the diffusion step 4) depends on the concentrations $\theta_i(0)$ and $\theta_i(l)$ of interstitials at the phase boundaries of the hydride and is a function of the thickness and shape of the sample. If the hydride formation step 5 is described as a phase transition of hydrogen atoms from the hydride to the reaction front at the interphase hydride/metal the rate equations are the same as for step 3.

Table 2.2 Rate equations and equilibrium constants of partial steps (2.20)

physisorption	chemisorption	surface penetration
$v_{1+} = k_{1+}p(1 - \theta_p)$	$v_{2+} = k_{2+}\theta_p(1 - \theta_c)^2$	$v_{3+} = k_{3+}\theta_c(1 - \theta_i)$
$v_{1-} = k_{1-}\theta_p$	$v_{2-} = k_{2-}\theta_c^2(1 - \theta_p)$	$v_{3-} = k_{3-}\theta_i(1 - \theta_c)$
$\dfrac{\theta_p}{p(1-\theta_p)} = K_1$	$\dfrac{\theta_c^2(1-\theta_p)}{\theta_p(1-\theta_c)^2} = K_2$	$\dfrac{\theta_i(1-\theta_c)}{\theta_c(1-\theta_i)} = K_3$

2.2.3 Combining the Partial Steps

The interrelation between individual partial steps of a reaction is accomplished by the concentrations of reacting species which are increased or decreased at specific local areas of the system. In the present model the following areas and species have to be considered (Table 2.2):

i) MH_2-surface

$$\frac{d\theta_{H_2(phys)}}{dt} = v_{1+} + v_{2-} - v_{1-} - v_{2+}, \tag{2.27}$$

$$\frac{d\theta_{H(chem)}}{dt} = v_{2+} + v_{3-} - v_{2-} - v_{3+}; \tag{2.28}$$

ii) subsurface layer

$$\frac{d\theta_{H(in\ MH_2)}}{dt} = v_{3+} - v_{3-} - I_{diffusion}; \tag{2.29}$$

iii) MH_2/M-interface

$$\frac{d\theta_{H(at\ interface)}}{dt} = I_{diffusion} - v_{5+} + v_{5-}. \tag{2.30}$$

In addition to the rates of the partial reactions, the equation for the diffusion flux through the hydride coating is needed which can be approximated by the relation

$$I_{diffusion} = D_{H(in\ MH_2)} \frac{c_{H(surf)} - c_{H(MH_2/M)}}{l} \tag{2.31}$$

with l hydride layer thickness.

2.2.4 Mathematical Solution of the Problem

The final aim of modeling is the determination of a rate law, which gives the growth rate of the MH_2 layer

$$dl/dt = v = v(p_{H_2}, T)$$

and a time law which gives the layer thickness

$$l = \int_0^t v\, dt = l(p_{H_2}, T, t).$$

The rate constants k of the partial equations and the diffusion coefficient D shall be given and the experimental parameters pressure p_{H_2} and the temperature T shall have fixed values. However, the concentrations θ or c of the species $H_2(phys)$, $H(chem)$, $H(subsurface)$, and $H(in\ MH_2)$ are not known. They can be evaluated by solution of the four differential equations (2.27–30). Unfortunately, this kind of an exact and comprehensive solution cannot be achieved by a simple method. The general result would include not only the rate and time laws in steady state but also the incubation periods for a system starting from an arbitrarily chosen initial state until steady-state conditions are established at constant pressure and temperature.

2.2.5 Steady-State Conception

At the begining of a chemical reaction, the concentration values of the intermediate species are normally not exactly defined. But after a relatively short period of time, a steady state is reached and the concentrations have attained such values which ensure that the flux of atomic or ionic species is the same at each point along the reaction path in the system. The change of local concentrations is in almost all cases of practical interest small and can be neglected. Small means the flux of species passing across the sample on the way to the reaction front is much higher than the amount of particles needed to adapt the concentration profile to the slowly changing conditions of the steady state. In this limit where all changes of intermediate species in (2.27–30) are assumed to be zero, the set of differential equations can be replaced by a set of algebraic equations. This simplifies the mathematical problem enormously and the accuracy of results is not reduced as long as most hydrogen atoms are transfered through the hydride layer to the reaction front. The amount of material consumed by concentration changes inside the sample can be easily estimated as well as the small errors arising from the use of this extremely helpful approximation (Sect. 5.4.2).

2.2.6 Rate Determining Step

Even after introduction of the steady state conception, the general solution of the set of algebraic equations remains still a serious mathematical problem. Therefore, most reaction kinetic treatments and models use, in addition, a much more simplifying conception, namely, the assumption of only one rate determining step. The limiting cases obtained after introduction of this efficient, but very restricting, approximation provide a reliable quantitative description of the reaction kinetics of a system only if a single partial reaction step is the bottleneck of the process considered under specific experimental conditions. It will be demonstrated in subsequent chapters that this restriction to the limiting cases frequently cannot afford a correct simulation of the data measured for real systems. In the present example diffusion of hydrogen atoms through the growing hydride layer is one of the partial steps. It becomes slower with increasing thickness. One has to compare this step with surface reactions, which are pressure dependent, but they are not affected by the reaction time or the layer thickness. Thus, it can easily happen that in the course of an exposure run the reaction is at first surface controlled and later on it becomes diffusion controlled.

The mathematical procedure for the calculation of rate and time laws in the approximation of a rate determining step is easy. The formulation of the expression for the rate law of the total reaction begins with the rate law for the forward reaction of the rate determining step. The reaction rate is given there as a function of concentrations of intermediate species. These concentration terms must be replaced by the pressure as an experimental parameter using the equilibrium constants for the preceding reaction partial steps given in Table 2.2. In the present

Table 2.3 Rate and time laws for different rate determining steps of the hydride formation reaction

partial step	rate law		time law
Physisorption	$dl/dt = kp$	$k = k_{1+}$	$l = kpt$
Chemisorption	$dl/dt = kp$	$k = K_1 k_{2+}$	$l = kpt$
surface penetration	$dl/dt = kp^{1/2}$	$k = (K_1 K_2)^{1/2} k_{3+}$	$l = kp^{1/2} t$
Diffusion	$dl/dt = k' \Delta\theta(p)/l$	$k' \Delta\theta_i \approx kp^{1/2}$	$l = kp^{1/2} t^{1/2}$
hydride formation	$dl/dt = k(p)$	$k = k_5 \theta_i(p)$	$l = k(p)t$

approach it is assumed that the forward and the backward reaction rate of all steps preceding the rate determinig step are much higher than the small reaction rate at the rate determinig step. Thus, the assumption of quasi-equilibrium for the partial processes before and after the rate determining step is justified. Table 2.3 shows rate and time laws obtained for the limit that only one of the partial steps is rate determining and that in addition the terms $(1 - \theta)$ for the concentration of empty sites in the equations of Table 2.2 can be set equal to one.

2.2.7 Concluding Remarks

The structures and mathematical treatment of all models on reaction mechanisms and kinetics are based on the interaction of individual partial reactions and of transport mechanisms for migrating species. Therefore, the same principles and procedures have to be applied as outlined for the simple example of hydride scale formation on a plane metal sheet under ideal conditions. The structure of a model is defined by the species introduced, the reaction equations of partial reactions and the equations for transport mechanisms. Each of the intermediate species occuring in the model needs a reaction for formation and one for annihilation. The number of concentration terms for intermediate species introduced must be equal to the sum of equations resulting from the steady state conditions (2.27–30), the expressions for fluxes and/or other independent equations containing concentration terms. This is a necessary condition for the solution of the mathematical problem and simultaneously also a proof that the conception of the model is based on reliable assumptions.

The main difficulty of modeling is the mathematical treatment. Very frequently further approximations have to be introduced and sometimes it is not easy to keep the improvement in mathematics and the loss of realistic assumptions in fair balance. The understanding of a reaction mechanism can be impeded more than improved by a more advanced model if it introduces unrealistic correlations for additional partial processes and boundary conditions.

A model describing a reaction mechanism should be able in any case to present calculated rate and time laws which can be checked by experimental curves. Otherwise the efforts of the author are useless for a reader who wants to understand his findings. Furthermore, the parameters of a model should not merely be arbitrary fitting parameters but have a clear physical significance. It should be possible to compare the order of magnitude of the values needed for curve fitting at least qualitatively with literature data or appropriate estimates.

2.3 Characteristics of Reaction Partial Steps

In most models of a complex metal-gas reaction relatively simple chemical reactions and transport mechanisms can be chosen as partial steps. They simulate the main features of the process of interest with a small number of species as intermediate reaction products. If too many parameters are introduced into a model, fitting of experimental data is no problem but the interpretation of the phenomena becomes less conclusive. Each partial step has its inherent structure and specific limitations. This should be kept in mind when it is used in a model. The values of constants and parameters in rate equations and equilibrium constants must be of the correct order of magnitude according to their physical meaning defined by the partial step; otherwise no atomistic model of the reaction mechanism has been developed but only a mathematical structure suitable for the fitting of curves.

The distinction between reasonable and unreasonable assumptions for a possible reaction mechanism and the discussion of results is often facilitated if relative units are used for concentrations and fluxes which give a clear idea what is meant instead of SI units. The transformation from relative units to the SI units in the final result calculated for a specific system is no problem. A relative unit frequently used in surface science and in this book is the atomic monolayer (ML) which is the number of atoms per m^2 of a surface or a lattice plane. For subsequent discussions one should remember the order of magnitude of some basic quantities of solid state physics.

Number of atoms/m^3 in condensed state: $c_{max} \approx 10^{28}/m^3$
Number of atoms/m^2 on a lattice plane: $1\ ML \approx 10^{19}/m^2$
Frequencyy of lattice vibrations: $v \approx 10^{13}/s$
Distance between lattice planes: $a \approx 0.3$ nm
Impinging rate of gas molecules: $r^0 \approx 10^9 \cdot p[bar]$ ML/s

If a model can be described by a plane geometry, only one local variable has to be considered and the mathematical treatment is simplified enormously. In such cases, it is very convenient to give local concentrations in terms of coverage or occupancy θ, $0 < \theta < 1$, of available sites of a species in lattice planes perpendicular to the reaction path. The unit for fluxes can be given in monolayer equivalents per second: $1\ ML/s \approx 10^{19}$ species/$m^2 s$. A flux of one ML/s forms about one lattice plane of oxide per second or about one μm/h. The use of the relative quantities monolayer (ML) lattice distance a, and coverage θ, for the characterization of areal densities of a species on surfaces, interfaces, and lattice planes, does not affect the mathematical treatment of the models. If experimental results and simulated curves have to be compared for a real system, the quantities ML and a can be defined for this special problem and inserted into equations or the scales of graphs obtained by numeric methods. Relevant definitions and data are given in Appendix E.

2.3.1 Molecular Adsorption or Physisorption

The molecular adsorption or physisorption of a gas molecule G_2 on a surface is accomplished by weak van der Waals or dispersion forces. The binding forces are less than 20 kJ per mol and the molecules are mobile on the surface. Even this most simple type of an adsorption process is a rather complex phenomenon when studied experimentally or theoretically in more detail [2.11–14]. However, for our purpose this process shall be reduced to the essentials [2.10]. The reaction equation and the rate equations can then be written as

$$G_2(\text{gas}) + \text{site} \leftrightarrows G_2(\text{phys}) \tag{2.32}$$

and

$$v_1 = k_{1+}p_{G_2}(1 - \theta_{\text{phys}}), \tag{2.33}$$

$$v_{1-} = k_{1-}\theta_{\text{phys}}. \tag{2.34}$$

The rate constant k_{1+} is the impinging rate of gas molecules

$$k_{1+} = \alpha \cdot \frac{2.7 \cdot 10^{29}}{\sqrt{MT}} \ [\text{K}^{1/2}\text{bar}^{-1}\text{m}^{-2}\text{s}^{-1}]. \tag{2.35}$$

The condensation coefficient α is usually unity or in the range 0.1 to 1. For molecular weights M of about 30 and $T \approx 300$ K k_{1+} is about 0.3×10^9. The rate constant k_{1-} is the probability that a gas molecule physisorbed in the shallow potential could (Fig. 2.3) escape back to the gas phase. This constant is proportional to the frequency factor $v \approx 10^{13}$/s times the Boltzmann factor $\exp(-\Delta H/kT)$. Here, ΔH is the negative quantity of the physisorption energy ΔH_{phys} and acts as an activation energy term since no energy barrier exists between the gas phase and the physisorbed state. The equilibrium condition $v_+ = v_-$ is given as

$$\frac{\theta_{\text{phys}}}{(1 - \theta_{\text{phys}})} = p \cdot K_p(T) \approx p[bar] \cdot 10^{-4} \cdot \exp(-\Delta H/kT) \tag{2.36}$$

or

$$\theta_{\text{phys}} = \frac{p \cdot K_p(T)}{1 + p \cdot K_p(T)}. \tag{2.37}$$

These relations show that θ_{phys} is proportional to the gas pressure. The term $(1 - \theta_{\text{phys}})$ represents empty sites and guaranties that θ_{phys} cannot exceed one monolayer at high pressure.

The low value of the physisorption enthalpy of about −20 kJ/mol, which can rise to −40 kJ/mol in some special systems, has the consequence that physisorption disappears almost totally at higher temperatures (Fig. 2.1). Therefore, models on the kinetics of metal-gas interactions at temperatures higher than 1000 K should not use physisorption as an intermediate state on the surface. The real situation is described much better by a collision process where the impinging molecules are either reacted or reflected.

Precursor state
The coverage of a surface with molecules in the physisorbed state can be rather high at ambient or low temperatures and because of the mobility of physisorbed species the rate of subsequent reactions is increased remarkably. The activation

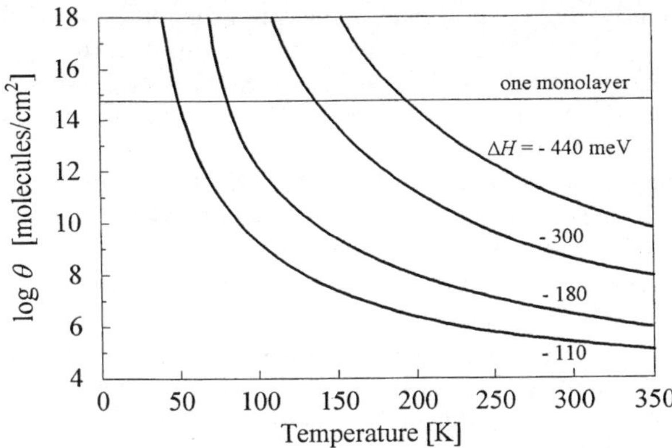

Fig. 2.1 Coverage θ_{phys} of the physisorbed state as a function of the physisorption energy ΔH (m = 28, $p = 10^{-4}$ mbar)

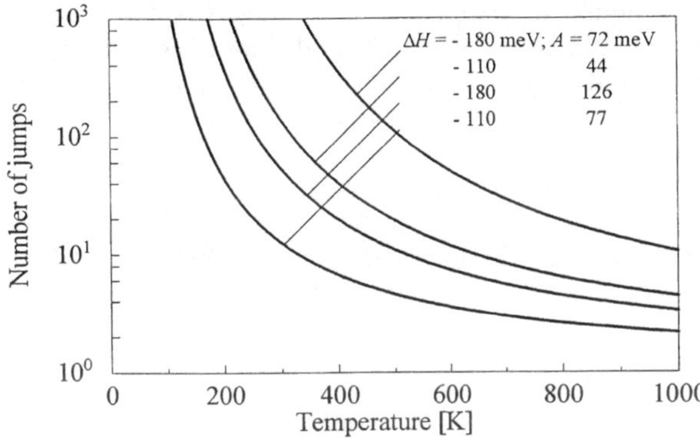

Fig. 2.2 Number of surface diffusion jumps of a physisorbed molecule before desorption

energy of diffusion jumps of physisorbed molecules is smaller than the adsorption energy. Therefore, the probability that a molecule jumps to a neighboring site is higher than for the desorption process. The relative probability for a diffusion jump compared to desorption or the number Z of diffusion jumps a molecule will perform before it is desorbed is given by

$$Z = \exp\left[(\Delta H_{\text{phys}} - A_{\text{diff}})/kT\right] \tag{2.38}$$

with ΔH_{phys} physisorption energy and A_{diff} activation energy of surface diffusion. Figure 2.2 shows that at temperatures near or below ambient temperature adsorbed molecules can jump to many neighboring sites before they leave the surface. This process affects strongly the kinetics of low temperature metal-gas reactions. If the incident gas molecule is adsorbed at first on a surface site which is already oxidized or inactive for other reasons, it jumps to several other sites in the neighborhood before it desorbes. On its way it will probably find a reactive site, if there is any, and undergo a surface reaction. The mobile physisorbed molecules act as a precursor state which enormously increases the probability for a subsequent chemisorption process. Consequently, the reaction rate or sticking probability, r, of an irreversible reaction or of a process which is far from equilibrium does not decrease proportional to the relative number of free surface sites, $(1 - \theta_{phys})$, but it shows an extended plateau near one and the decrease of the reaction rate occurs only at θ values close to one (Fig. 5.7).

Numerous models have been developed to describe this effect in more detail [2.15, 16]. In models where this aspect of the mechanism is not of specific interest a function $(1 - \theta)^n$ with n > 1 instead of $(1 - \theta)$ in the Langmuir equation (2.36) yield r vs. θ curves for chemisorption reactions with a plateau-like shape very similar to the results of more advanced precursor models. The introduction of merely an exponent n can be justified if the forward reaction v_{1+} is assumed to fail only when more than n neighboring sites next to the site of impingement are already reacted. At low coverages the equilibrium conditions (2.36, 37) can be approximated by the relation $\theta \sim p$.

In almost all models on kinetics of low-temperature metal-gas interactions physisorbed species are used as intermediate species for the subsequent partial reactions in condensed state. If the impingement rate is higher than the overall reaction rate, or in other words, if the reaction probability of impinging gas molecules is less than 10^{-2} it can be assumed that a physisorption equilibrium is established. In most models physisorption is the only partial step which contains the gas pressure as an explicit variable. For biatomic gas molecules such as O_2, N_2 or H_2 this has the consequence that a pressure dependence $v \sim p$ is found only if the rate determining step is either gas phase transport at very low pressures or dissociation of gas molecules physisorbed on the sample surface under quasi-equilibrium conditions. If other partial steps, for example, diffusion processes or surface penetration, affect the reaction rate as slow partial processes the rate law has the form $v \sim p^n$ with n being a factor between 0.5 and 1.

2.3.2 Chemisorption.

The atoms of adsorbed gas molecules can form chemical bonds with the atoms of the metal substrate. This new state is called chemisorption. The binding energies are of the same order of magnitude as in chemical compounds and attain values up to 600 kJ or 6.5 eV per g atom oxygen on reactive metal surfaces. The surface diffusion rates of chemisorbed species can be compared with grain boundary diffusion in solids. But at low temperature chemisorbed atoms are usually considered to be immobile.

A detailed analysis of chemisorption processes on an atomistic scale touches almost any field of surface science. The bond energies and activation energies of the surface reactions depend on the lattice structure and defect structure of the surface. Active sites for a chemisorption reaction can be normal sites, steps, adatoms, dislocation cores or impuritiy atoms. The electronic structure of surface atoms is different from the bulk state. Thus, also electronic surface states are of specific interest. Even the successful theorem of steady state conditions in kinetics does not hold for some catalytic surface processes. If reconstruction of the surface, which means a change of the structure, is caused above a certain adsorbate concentration periodic fluctuations with time are observed in the reaction rate [2.17]. An introduction to the phenomena of surfaces and interfaces which covers also the experimental techniques available has been recently published by *Lüth* [2.18].

Theoretical aspects have been presented in an overview of *Desjonque'res* and *Spanjaard* [2.14] and papers on reversible chemisorption of gases on metal surfaces as typic processes in catalysis are found in the book of *Madix* [2.19]. The mechanism of chemisorption reactions has been described by *Ehrlich* [2.15] and further relevant papers are published in [2.16].

For modeling one has to reduce the phenomenon chemisorption to the few essentials really needed. A chemisorbed species shall be a particle which has been formed at the substrate surface by a chemical reaction. The reactants can be impinging or physisorbed gas molecules, electrons, and atoms or defects on regular or specific sites of the substrate. The chemisorption species can have an electric charge and consist of one or of several atoms. Typical chemisorption species on an oxide surface are: $O_2, O_2^-, O_2^{--}, O, O^-$ and O^{--}.

The definition of a chemisorption species as a charged point defect on the surface can be considered also as an approximation in a model for a real situation where in a localized area of the surface a cluster is formed with excess of one or two oxygen atoms and zero to two electrons when compared with the mean composition of the oxide layer. In models on reaction kinetics the interactions between atomic, molecular and ionic species are treated as chemical reactions. Therefore, the surface states must be described by chemisorption species which are produced or decomposed by chemical reactions and characterized by a concentration at the surface. This definition does not prevent principally model structures, as discussed in the next chapters, if more complex surface defects have to be considered. Most surface states can be simulated by a set of properly chosen surface defects which interact by chemical reactions in such a way that the real situation is represented adequately.

The presence of charged chemisorption species on the surfaces of oxide layers is a very important precondition for the models on low-temperature oxidation. They provide acceptor sites for the electron transfer from the metal phase to the oxide surface. The energy levels of electronic surface states can lie within the energy gap present inside the oxide. The number of papers on the electronic structure of oxide surfaces is much smaller than for metal surfaces or for semiconductors [2.20] but the present understanding of the phenomena [2.21–24] does not raise serious objections against the present definitions for charged and uncharged chemisorption species.

Most models on reaction mechanisms work properly if only one chemisorption species is introduced as an intermediate surface particle which acts simultaneously as electron acceptor for excess surface charges. The electric field produced across oxide layers or other semiconducting coatings is necessary to guarantee charge neutrality if charged particles are migrating. The use of only one chemisorption particle is no severe restriction for a model as long as the concentrations of other possible candidates are assumed to be smaller by one or more orders of magnitude.

Energetics
The energy values for the equlibrium states and the activation barriers of mobile species introduced in a model are depicted best by diagrams reaction path vs. energy. Energies are normally given as enthalpy values, ΔH, since systems are usually considered under isothermal isobaric conditions where the minimum of the Gibbs free energy $G = H - TS$ defines the equilibrium state. Figure 2.3 shows

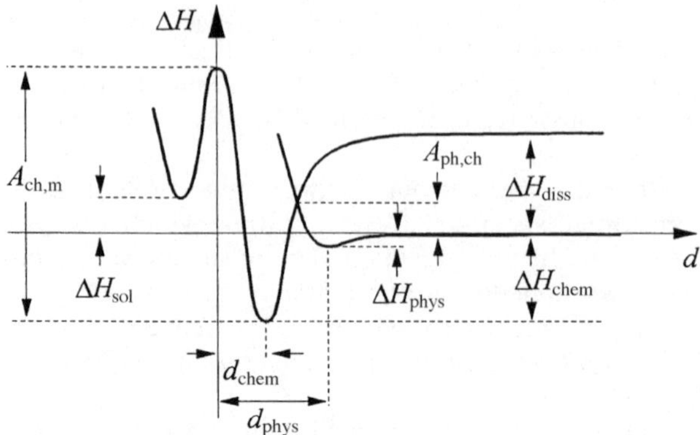

Fig. 2.3 Energy states for physisorbed molecules and chemisorbed atoms on a surface

a diagram for a bimolecular gas species, for example, O_2 or H_2. The energy of the gas molecule in the gas phase is chosen as reference state for the the enthalpy differences indicated. If a gas molecule approaches the metal surface, it is trapped at first in the shallow energy minimum of physisorption, ΔH_{phys}, at the distance d_{phys}. Since molecules cannot penetrate the metal surface the potential curve increases strongly with decreasing distance from the surface.

The second curve in Fig. 2.3 represents energy states of the atoms after dissociation. The curves begin at the right-hand side at the positive level of the dissociation energy ΔH_{diss}. If a stable chemisorption state of the gas atom exists on the surface, a potential minimum is found with a negative chemisorption energy ΔH_{chem}. Gas atoms can also be dissolved in the metal, for example, at an interstitial site. The energy of an atom in this state is given by the heat of solution, ΔH_{sol}. The ΔH values of the chemisorption and solution energy are given for one atom and not for one molecule. It is an inherent problem of energy versus reaction-path diagrams that activation energies have to be defined exactly for the number of particles forming the activated complex. During dissociation the gas molecule is lifted at first to the level of the activation barrier $A_{\mathrm{ph,ch}}$ and then two atoms drop into the energy minimum of chemisorption. In the reverse recombination step, two chemisorbed atoms loose the chemisorption energy and need additional energy to reach the activation barrier. These two processes are visualized best in a diagram where the chemisorption energy is counted for two atoms. However, in the forward and backward reaction of surface penetration only one atom is involved and, thus, all energies must be given per atom, especially for the activation barrier between the two equilibrium states, $A_{\mathrm{ch,ss}}$. This demonstrates that the optimum choice for the definition of the chemisorption energy depends on the direction of the partial step under consideration.

A very important detail in Fig. 2.3 is the intersection of the two curves for the O_2 molecules and the O atoms in the surface region. It demonstrates that the activation energy for the dissociation of a physisorbed gas molecule on a surface can have values between zero and the dissociation energy ΔH_{diss}. Zero values are frequently observed when oxygen or other highly reactive gases are absorbed by metals. In less reactive systems the activation energies are in the range of 50% of the dissociation energy, ΔH_{diss}, or higher.

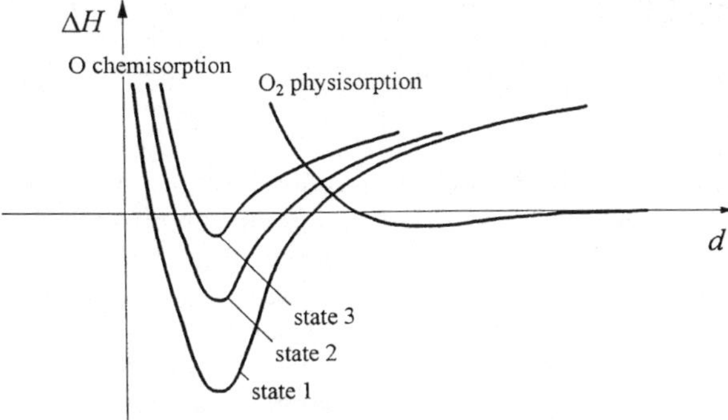

Fig. 2.4 System with different chemisorption sites

It has been previously mentioned that physisorbed oxygen molecules can be chemisorbed at different surface sites and form species with different negative charges. Consequently, also the chemisorption energy levels must be different. This is schematically shown in Fig. 2.4. Higher chemisorption energy levels, which means weaker chemisorption bonds, shift the activation energies of dissociation to higher values. The more stable sites at low energy levels are, therefore, occupied first in an absorption process with relative fast rates. With increasing coverage of a surface less stable sites are filled. Even on an ideal surface without defects such as edges, steps, vacancies or surface ad-atoms the chemisorption energy can decrease with increasing coverage due to repulsive interaction of neighboring chemisorption species. The decrease of the chemisorption energy with increasing coverage has been proved by flash desorption experiments [2.16, 25] and [Ref. 2.18, pp. 456–471]. If the temperature of a sample is increased in a vacuum receiver with constant heating rate, normally not only one pressure peak is observed during gas desorption but several of them. This demonstrates that more than one chemisorption states are present on the surface. At low coverage of the surface the peaks at low temperature disappear since only the most stable state is occupied.

Rate laws and equilibria
The reaction equation for the formation of charged or uncharged O_2^{z-} oxygen chemisorption species from a physisorbed O_2 molecule can be written as

$$O_2(\text{phys}) + \frac{2}{x}\,\text{sites(chem)} + \frac{2z}{x}\,e \leftrightarrows \frac{2}{x}O_x^{z-}(\text{chem}) + \text{site(phys)} \tag{2.39}$$

with $x = 1$ or 2 and $z = 0, 1$ or 2. If the chemisorption species is an uncharged oxygen atom and the unoccupied sites are written as $(1 - \theta_{\text{chem}})$ the reaction rates are

$$v_{2+} = k_{2+} \cdot \theta_{\text{phys}}(1 - \theta_{\text{chem}})^2, \tag{2.40}$$

$$v_{2-} = k_{2-} \cdot \theta_{\text{chem}}^2(1 - \theta_{\text{phys}}), \tag{2.41}$$

and the equilibrium condition reads as

$$\frac{\theta_{\text{chem}}^2 \cdot (1 - \theta_{\text{phys}})}{(1 - \theta_{\text{chem}})^2 \cdot \theta_{\text{phys}}} = K(T). \tag{2.42}$$

We are reminded here once again that in a correct rate equation the number of reactants must be exactly the number of species which are really involved in the atomistic process and not half or twice of it. This is different from equilibrium considerations. If the reactants and the energy terms in equations for equilibria are multiplied by the same factor, the results remain correct but rate laws become totally wrong. One really needs, for example, exactly one physisorbed O_2 molecule and two empty sites for the formation of two chemisorption particles at the reaction site on the surface and the probability that the reaction really takes place increases with temperature proportional to $\exp(-A/kT)$ and not proportional to $\exp(-nA/kT)$ where n is a constant multiplication factor.

If charged particles are involved in a reaction the situation becomes more complex. An electric field existing parallel to the direction of the particle transfer along the reaction path affects the activation energy by an additional term $z_i \cdot e \cdot E \cdot a/2$ where z_i is the charge of the particle and $a/2$ the distance of the activation barrier from the initial site of the charged particle,

$$A' = A + z_i e\, E\, a/2. \tag{2.43}$$

Figure D.1 shows how activation energies are affected in the presence of an electric field. The equilibrium constant $K = \exp(-\Delta G^0/kT)$ is extended by a field term, too, and is given for an electrochemical reaction by an expression

$$K' = \exp\big[-(\Delta G^0 + z_i e \cdot E \cdot a)/kT\big]. \tag{2.44}$$

2.3.3 Formation of Lattice Defects

In solid-state reactions where dense surface layers are formed at least one of the reacting elements has to migrate as lattice defect through the layer to the reaction front. The mobile defects are formed by interface reactions either at the surface or at the interface reaction-product/substrate. The most simple example of point defect production at the surface of a material is the transition of a chemisorbed uncharged atom to an interstitial site in the first subsurface layer of the bulk.

$$O_{\text{chem}} \leftrightarrows O_{\text{sol}}, \tag{2.45}$$

$$\begin{aligned}
v_{3+} &= \theta_{\text{chem}} k_{3+} \exp(-A_{3+}/kT), \\
v_{3-} &= \theta_{\text{sol}} k_{3-} \exp(-A_{3-}/kT),
\end{aligned} \tag{2.46}$$

$$\frac{\theta_{\text{sol}}}{\theta_{\text{chem}}} = K_3(T). \tag{2.47}$$

This process depicted schematically in Fig. 2.3 is called *segregation* reaction. The formation of an interstitial metal atom in an oxide at the metal/oxide interface yields almost identical equations.

The formation of charged defects requires reaction equations which guarantee the conservation of mass and charges. An O^{--} interstitial in an oxide can be formed, for example, from O_2^- chemisorption species according to the reaction

$$4O_2^-(\text{chem}) \leftrightarrows 2O^{--}(\text{sol}) + O_2(\text{gas}) \tag{2.48}$$

If electric fields have to be considered in reaction models with charged particles the activation energy and the equilibrium constant are affected and contain potential terms (Appendix A.2).

2.3.4 Formation of the Reaction Product

The reaction product of a solid state reaction is normally formed at an interface which is called the *reaction front*. Part of the reactants are already present there and the other migrate as lattice defects through the layer. For example, the formation of an oxide lattice molecule at the oxide/metal interface can be written as

$$O(\text{interstitial}) + M(\text{bulk}) \leftrightarrows MO \tag{2.49}$$

or at the oxide surface

$$M(\text{interstitial}) + O(\text{chem}) \leftrightarrows MO. \tag{2.50}$$

The reaction equations chosen again have to obey the laws of mass conservation and charge neutrality.

2.3.5 Diffusion Processes

The migration of reacting species through solid phases is an indispensable partial step if adherent layers are formed on a substrate. The most simple transport mechanism is diffusion of uncharged particles according to Fick's first law

$$J = -D \cdot \text{grad}\, c. \tag{2.51}$$

If the layer growth dl/dt is assumed to be proportional to the flux J the rate law becomes

$$dl/dt = R \cdot D \cdot (c_{\max} - c_{\min})/l \tag{2.52}$$

and by integration the oxidation time law is obtained which is also known as Tammann's law

$$l = \sqrt{2R(c_{\max} - c_{\min})} \cdot \sqrt{D \cdot t}. \tag{2.53}$$

This solution of a diffusion-controlled layer-growth mechanism on a plane infinite surface demonstrates the general features of diffusion-controlled processes. If the diffusion constant D of the migrating particle does not depend on concentration, an assumption which is obeyed for ideal diluted systems, the concentration of dissolved particles in a matrix can be presented in plots $c(t)/c_{\max}$ versus $x/\sqrt{D \cdot t}$ with x being the distance from the plane where the diffusion flux enters the bulk [2.26]. Typical standard solutions of diffusion problems are shown in Figs. 2.5–2.7. In Fig. 2.5 the concentration profile in the bulk of a semi-infinite plane is

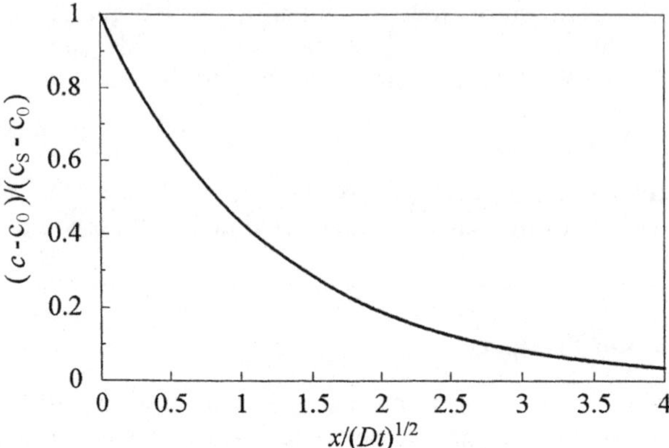

Fig. 2.5 Concentration profile in an infinite plane sample with constant surface concentration c_s and initial concentration c_0

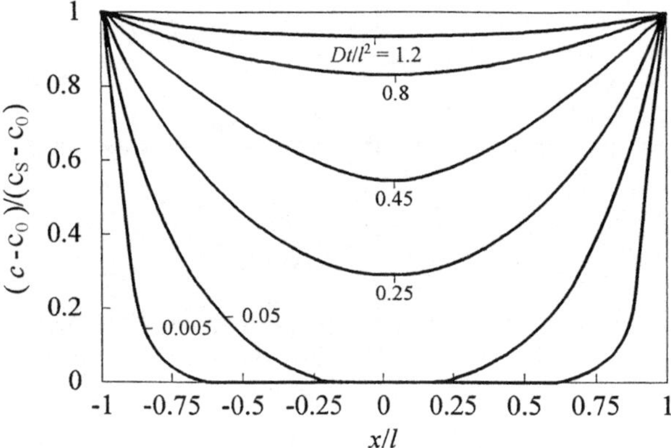

Fig. 2.6 Concentration profile in a sheet of thickness $2l$ for a diffusion flux from the surface to the bulk

shown in relative units $(c - c_0)/(c_s - c_0)$ as a function of the parameter $x/\sqrt{D \cdot t}$. The initial concentraion in the bulk is c_0 and c_s is the constant surface concentration. It represents the function

$$\frac{(c - c_0)}{(c_s - c_0)} = 1 - erf\left(\frac{x}{2 \cdot \sqrt{D \cdot t}}\right). \tag{2.54}$$

This curve can be used for estimates on the thickness of surface diffusion layers if the diffusion constant D is known. The thickness is about $l \approx \sqrt{D \cdot t}$ after an exposure time of t seconds.

The concentration of a species in a sheet can be increased by diffusion from outside. The concentration profiles developed in the sheet after increasing exposure times are shown in Fig. 2.6 as a function of the normalized time parameter $t \cdot D/l^2$. The thickness of the sheet is $2l$. The concentration in the sheet has reached the final value $\bar{c} \approx c_s$ after a period of time of about $t = l^2/D$. The speci-

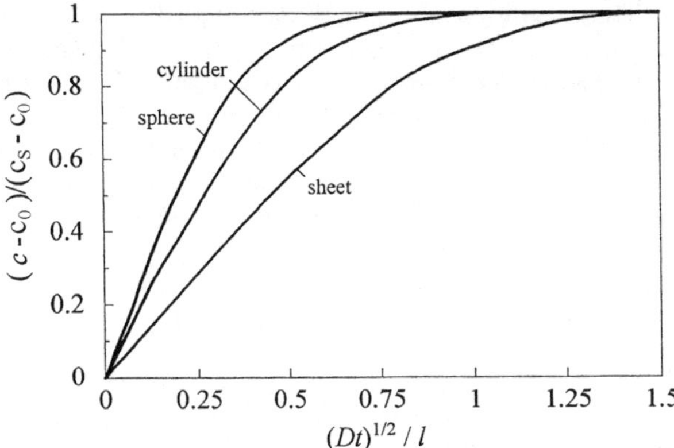

Fig. 2.7 Average concentration in a sheet (thickness $2l$), a wire, and a sphere (radius l)

fic shape of a sample does not drastically change the results of diffusion controlled processes. This can be seen from the curves in Fig. 2.7. There the ratio of the increase of the average concentration $(\bar{c} - c_0)$ to the final change of the sample concentration $(c_s - c_0)$ is plotted versus time in units of the parameter $\sqrt{t \cdot D}/l$ for a sheet (thickness $2\,l$), a wire (radius l) and a sphere (radius l). The time needed to obtain 50% saturation differs only by a factor of six that means by less than one order of magnitude.

The relation

$$l^2 = t\,D \tag{2.55}$$

is a very useful equation for the estimation of the thickness of diffusion layers from diffusion data. It shows, for instance, that for $D \approx 10^{-10}$ cm^2/s the diffusion zone after one hour is about 5 μm thick and after one year about 0.5 mm.

At the end of this section it should be mentioned that diffusion control is very frequently the rate determining partial step of solid state reactions, but it is not always the case. This is occasionally not noticed in papers presenting data on reaction kinetics. Whether or not diffusion control is responsible for the reaction mechanism can be tested in most cases by comparing experimental results with the rate and time laws of models where diffusion is assumed to be the rate determining step. The pressure and concentration dependence in the expressions are normally quite different for diffusion control and for surface control.

2.3.6 Electronic Currents

The defects responsible for the transport of reacting chemical elements through semiconducting reaction products are in most cases ions. Consequently, the charge distribution in the system is changed by the fluxes. If the equivalent of one mono-layer (10^{19} atoms/m^2) is moved as ions from one end of a 10 nm thick oxide scale to the other, its thickness is increased only by one lattice plane or 0.3 nm. However, the electric potential difference across the layer caused by the charge transfer would be in the range of 100 V. This demonstrates that in real systems the flux of charged particles has to be compensated and the net charge transfer of all fluxes must be zero in steady state. This condition is normally accomplished by additional electronic fluxes. In thicker semiconducting layers positive and negative charge fluxes are coupled by an electric field which enhances slightly the slower flux and retards the faster one, if necessary by orders of magnitude.

Very thin semiconducting layers with a thickness of about one nm or less can be penetrated by tunneling electrons. Since the tunnel current decreases exponentially with the layer thickness, the transport of electrons through thicker scales must be accomplished by a hopping mechanism or by diffusion processes where electrons have to be transfered to the conduction band of the semiconducting layer. Therefore, all models which consider electron tunneling as a possible transport mechanism in the initial stage of oxidation must contain at least one additional transport mechanism for electrons which sustains the electronic current through the layer at the later stages of the process. More detailed informations on electronic currents are found in Appendix C.

2.3.7 Ionic Fluxes

The driving forces for the migration of ionic defects through a semiconducting surface layer are the gradient of the chemical potential, which means a concentration gradient, and electric fields. High electric fields can exist in very thin layers as a consequence of surface charges. This so-called Mott field is created, for example, by tunneling electrons which establish an equilibrium between electronic states in the metal and on the oxide surface (Appendix B). It reduces the effective activation barrier of migrating ionic lattice defects to very low values.

Other fields in semiconducting surface layers are caused by electrochemical quasi-equilibria of fast diffusing charged species. They are relatively weak and adapt the fluxes of the faster particles to the value of the slowest one. This result of the coupled current condition guarantees charge neutrality. The fluxes of slowly migrating species cannot be enhanced remarkably by diffusion potentials, only the faster ones are slowed down. Flux equations for charged species which hold for specific approximations are compiled in Appendix D.

2.3.8 Initial Stage of Oxidation

Experiments on the initial stage of oxidation or, more general speaking, of compound formation are normally performed with a clean metal sample on a well defined surface in an UHV receiver. The gas amount added in subsequent exposures is so small that the surface is at first not completely covered by chemisorbed atoms or by a thin oxide layer. The oxygen molecules absorbed can occupy different sites in the sample or on the surface, for example:
i)   They can be dissolved in the metal matrix and no adsorption layer is found;
ii) They can be chemisorbed as adatoms with a statistical distribution. The terms chemisorption and surface segregation are used to describe this behavior;
iii) They can be chemisorbed and form islands one monolayer thick if repulsive forces exist between the chemisorbed adatoms;
iv) The islands can be oxide phases more than one monolayer thick. If the surface and interface energies of the oxide are high and the surface energy of the metal low, then thick surface oxide particles are found. If the surface energies

are lower than the bulk energy plus the energy of the metal surface then the oxide phase spreads over the surface.

The equilibrium states of oxygen atoms in and on a metal phase can be defined by the following approach: The metal phase with all its defects such as point defects, impurity atoms, dislocations, grain boundaries and the surface is treated as a rigid system. The change of the Gibbs free enthalpy values shall be given if oxygen atoms from the gas phase are trapped in one of these defects states. The chemical potential of oxygen in this system is described as a function of concentration when the defect positions are successively filled starting from the sites with the lowest energy. This has the consequence that at very low concentrations where low energy sites are filled the simple relation of Henry's law is no longer valid. In classical thermochemistry it is assumed that only one site, normally an isolated interstitial site, is accessible for the oxygen atom. The error made by this approximation is for most practical problems neglectible since the amount of atoms trapped in defect sites is very small. The theoretical treatment of this kind of defect thermochemistry has been developed by *Kirchheim* and checked by experiments in metal hydrogen systems [2.27–30].

Segregation equilibria between the oxygen atoms dissolved in the bulk or trapped in defect sites and chemisorbed at the surface can be described by such models and simulation of the processes i to iii mentioned above can be accomplished without introduction of simplistic assumptions. The last process iv is more correctly described by nucleation theories for a second phase and terms for the surface and interface energies have to be considered in thermochemical calculations.

The correct theorectical analysis of the phenomena chemisorption, segregation, island formation and nucleation of oxide particles or oxide skins on surfaces is not trivial [Ref.2.18,p.72] since it deviates in some sensitive points from the traditional treatments of thermochemistry derived for the behavior of bulk atoms in the standard positions of ideal phases. Therefore, it is very important to check whether or not the definitions and equations used are suitably chosen for the characterization of the problem of interest. For the models on compound layer formation in reactive metal-gas systems discussed in subsequent chapters these processes are, fortunately, of minor practical relevance. If a clean reactive metal surface is exposed to oxygen or hydrogen and the gas supply is not artificially impeded in a UHV experiment, the surface is immediately covered by a dense surface layer. Thus, the surface reactions mentioned are normally finished within the incubation period before a steady state in the reaction mechanism is established. The more important question for model development is how the transport of the reactants, atoms or ions, through a chemisorption or compound layer to the reaction front can be accomplished in a reasonable way. Surface reactions which are really needed for the definition of a reaction mechanism are formation of chemisorption or defect species on the surface of the compound layers. These intermediate reaction products are required for linking the individual partial processes of a reaction model.

3 Experimental Techniques

Quantitative determination of the kinetics of low- temperature oxidation or of other reactions forming thin tarnishing layers on metal surfaces is a challenge for the investigator. The total amount of gas absorbed is less than 100 monolayer equivalents of gas molecules or 10^{14} to 10^{17} atoms per cm^2. This corresponds to 10^{-8} to 10^{-6} g/cm^2 or 10^{-5} to 10^{-3} cm^3 gas per cm^2 at 1 bar and 300 K. Thus, the sensitivity of methods for the determination of rate and time laws should be better than 10^{-8} g/cm^2 or 10^{-5} Ncm^3 /cm^2. Only hydrogen is absorbed in larger quantities due to its high mobility in the bulk. However, this reaction is normally suppressed or impeded by the oxide scale present on the surface. Therefore, reliable data can be measured only if the surface of the metal sample is not covered by poisoning elements from the beginning to the end of an experiment. Furthermore, varying the parameters pressure and sample temperature within reasonable limits should be possible. Unfortunately, no method enables the determination of low-temperature reaction kinetics in the same ideal manner as microbalance experiments do it in the field of high-temperature oxidation. Each one of the limited number of techniques applied so far has its specific advantages but also severe shortcomings. This chapter gives a short characterization of various experimental methods and their applicability to quantitative measurements will be discussed. The techniques are subdivided into the categories: volumetry, gravimetry, optics, high energy ion beams, X-rays, and surface analytical methods. All methods have one difficulty in common, namely the problem to define the initial state of the metal surface.

3.1 Initial State of the Metal Surface and UHV Experiments

At the beginning of an experiment the sample must have a clean surface. In high-temperature oxidation experiments an initial oxide scale of few nm thickness can be neglected since the growth of the oxide layer during an exposure run is normally in the mm range or more. However, quantitative measurements on low-temperature oxidation kinetics yield only then reliable results if the initial coverage of the metal surface is kept below one monolayer of oxygen atoms. The sticking probability of oxygen and water vapor molecules on a gas-free metal surface is close to unity. Therefore, preparation or cleaning of the sample surfaces must be performed under ultrahigh vacuum conditions at pressures below 10^{-6} mbar.

Once the problem of surface preparation in an UHV system has been solved, another difficulty emerges if the reaction mechanism shall be determined from reaction rate measurements as a function of time and pressure. In most papers presenting experimental results on the *initial stage* of low-temperature oxidation, a cleaned gas-free surface is exposed to small gas doses at very low pressures. The surface concentration of the samples is changed very slowly, for example, by a few monolayer equivalents within exposure times of some minutes up to several hours. Under such conditions the amount of absorbed gas molecules can exactly and continuously be monitored by the change of surface properties measured by the methods of low-energy electron diffraction (LEED), Auger electron emission (AES), emission of photo electrons (XPS, UPS) or secondary ion emission (SIMS) [3.1]. Valuable informations have been obtained by these techniques on the structure, energetics, or the valency of species on the sample surface for many systems but almost nothing can be said about kinetics. Since the impinging rate of the gas molecules has been reduced to extremely small values, the rate determining step of the overall reaction in reactive systems is automatically the transport of gas molecules in the gas phase. The only result of experiments on kinetics performed at extremely low pressure is the evaluation of the initial sticking probability, which is for oxygen molecules on clean metal surfaces normally one or close to one. If the oxide skin is growing on the surface, the sticking probability for oxygen molecules decreases drastically by many orders of magnitude. But then the sensitivity of surface analytical methods does not suffice to detect the small changes of surface properties which are far below one percent for an exposure of one Langmuir.

Attempts to avoid the trivial results of low-pressure experiments by gas exposures at higher constant pressures, for example at 10^{-4} to 10^{-2} mbar, creates other troubles. Lets assume it is possible to rise the pressure in a receiver from 10^{-9} mbar to a constant value at 10^{-4} mbar within only one second then the surface has been exposed to almost 100 Langmuirs or to the equivalent of about 100 monolayers of O_2 in this short period of time. This is in most metal-gas systems more than the total thickness of a stable oxide layer formed at ambient temperature. In such an experiment only the total amount of O_2 absorbed in this extremely short initial period of the experiment can be measured, but no rate or time law as a function of pressure and temperature. This unpleasant situation has at least one positive aspect for authors who propose models on the initial branch of metal oxidation kinetics. They cannot be disproved easily by experiment.

3.2 Volumetric and Manometric Methods

Volumetric and manometric techniques measure pressure changes of reacting gases in a closed volume. *Wagener* [3.2] has studied with this method the oxidation of thin films. At pressures below 10^{-3} mbar transport in the gas phase is accomplished by the molecular flow mechanism where collisions between gas molecules can be neglected when compared with collision with the walls of a receiver. The fluxes to a closed receiver can then be determined from the pressure difference measured across the ends of a capillary [3.2, 3]. The amount of gas molecules absorbed inside the receiver is obtained by integration over the exposure time. By this technique the small quantity of reactive gases absorbed at low temperature by clean metal surfaces can be measured quantitatively. However, two preconditions must be met. In a closed receiver only the metal sample should absorb gases but not any other part of the system and gas flows must enter and leave the receiver through capillaries or orifices with known conductivity L.

The gas flux J through a capillary to the receiver obeys the equation

$$J = L(p^0 - p^r) \tag{3.1}$$

with p^0 pressure in a gas reservoir on the outer end of the capillary and p^r the pressure measured in the receiver. Then, the absorption rate v can be written as

$$v = J - V_r \, dp/dt. \tag{3.2}$$

This means v is given by the flux J minus the gas amount required to increase the pressure in the receiver of the volume V_r. The amount of gas absorbed, N, is obtained from the integral of the reaction rate over the exposure time

$$N = \int_0^t v \, dt. \tag{3.3}$$

The conditions mentioned can largely be fulfilled if metal film samples are evaporated onto the walls of a glass sphere as a receiver. The sample covers more than 50% of the surface of a receiver consisting of an oxide which is inert with respect to oxidation reactions. Problems can arise from metal-gas interactions at the filaments of vacuum gauges. In an experiment the pressures in the receiver and in the reservoir volume are measured as a function of exposure time and the results can be presented as diagrams sticking probability s versus amount of absorbed gas, N. The pressure range of this technique is 10^{-10} to 10^{-3} mbar and the sticking probabilities can be determined from the maximum value 1 down to 10^{-5}. The minimum amount of gas detectable is less than 10^{-3} ML.

This method is also capable of measuring more complex metal gas interactions like poisoning processes in metal hydrogen systems if partial pressure gauges are used in the receiver and in the reservoir volume. Deposition of multilayer metal films on the wall of the receiver enables investigation of catalytic effects [3.4, 5] or hydrogen permeation measurements through the topmost film even if the material dissolves only traces of hydrogen at low pressure (Sects. 4.1, 6.1.2).

In the preceding section it has been mentioned that the kinetics of low-pressure gas exposure experiments performed with very small gas fluxes is mainly controlled by the transport of gas molecules in the gas phase with a rate law $v \propto p$. The pressure dependence of the reaction rate can be checked by repeated fast pressure changes after the initial stage of an experiment where the sticking coefficient is smaller than one. In most experiments with oxygen or with other reactive gases an initial plateau with a value of one or near one is found in plots sticking probability s versus absorbed gas amount, N. After the saturation concentration $N_{plateau}$ is reached the sticking probability decreases exponentially with further gas absorption. With reactions of more complex gas molecules sometimes only part of the atoms are absorbed, for example, oxygen atoms from H_2O molecules, and others are released to the gas phase. The term reaction probability, r, describes in this case the facts more precisely than the term sticking probability and is therefore preferred in subsequent chapters.

The most informative quantity obtained from volumetric experiments is the length of the plateau with r values near one in the r versus N plots (Fig. 5.7). For metal oxygen systems frequently it is very much longer than one monolayer equivalent of O_2, whereas for N_2 and CO it is about one monolayer. Another interesting finding is the fact that large amounts of O_2 can be absorbed also in a later stage of the oxidation reaction where the sticking probability is reduced from the initial plateau value one to 10^{-5} or less. Despite the lower reaction probability the reaction rate is still very high, even under high-vacuum conditions.

A big disadvantage of investigations with metal films is the undefined structure and roughness of the sample surface. Thus, the absolute values of the amount of absorbed gases measured for the geometric surface area of film samples is usually by a factor two to four higher than for massive samples. The installation of additional measuring systems for the characterization of the film surface structure in a receiver for volumetric experiments is problematic because they can undergo uncontrolled additional metal-gas reactions.

3.3 Quartz Crystal Microbalance

The quartz crystal microbalance technique is a gravimetric method with an extremely high sensitivity at ambient temperatures. It can be used for investigations on low-temperature oxidation or on other metal-gas interactions of film samples. The mass changes detectable are not as small as with the Wagener method but the sensitivity is with 10^{-1} monolayer equivalents still very high. Mass gain versus time curves can be measured from very short exposure times of a few seconds up to hours or days. An unpleasant limitation of the method is the fact that some time is needed to establish a constant pressure in a receiver. The pressure range of the method extends from ultrahigh vacuum to one bar or higher. Because of this very large volume the sticking or reaction probabilities can be measured in a single run from the maximum value one down to 10^{-13} with the same technique. This method is well suited for studies on low-temperature oxidation where very fast initial reaction rates are followed by very sluggish processes during more extended exposure times. Also, if the exposure pressure in isothermal isobaric experiments exceeds 10^{-7} mbar the kinetics of the initial branch of the reaction cannot be measured with the desirable accuracy by this technique.

The quartz crystal microbalance method takes advantage of the property of a quartz crystal to lower the resonance frequency if additional mass is deposited on the surface. The use of a quartz crystal as a tool for the measurement of the thickness of thin films has been described first by *Sauerbrey* [3.6]. An overview on the power of the method is given in the book edited by *Lu* and *Czanderna* [3.7]. For thin films the frequency shift is proportional to the mass deposited. Since the frequency of a 6 MHz crystal can be measured with the high accuracy of about $\pm\, 0.1$ Hz detecting mass changes equivalent to 0.2 monolayers of absorbed oxygen is possible. Problems can arise from the relatively large temperature dependence of the resonance frequency, which depends on the orientation of the cutting plane and the temperature of the quartz crystal. The maximum sensitivity is achieved only if experiments are performed close to ambient temperature. The temperature range of quantitative oxidation measurements can be extended to about 200 °C if the temperature of a quartz crystal, heated by infrared radiation, is carefully controlled [3.8].

The unpleasant problem appearing in all experiments with films cannot be avoided here also, namely, the roughness and not well defined structure of the sample surface. Since the mass of oxygen absorbed is measured per unit area of the geometric sample surface the real oxide layer thickness can be determined only if the roughness of the surface is known.

3.4 Ellipsometry

Another very valuable tool for studies of the kinetics of the formation of thin oxide films is the investigation of changes of polarized light after reflection from a sample surface. The use of this elegant technique has a long tradition [3.9–12]. It is very sensitive, and changes of the thickness of surface layers can be measured from less than one to more than ten thousand nanometers with high accuracy. In contrast to other methods the thickness of a film present on a surface before an exposure experiment can be determined if the optical constants of the substrate are known or can be measured with the ellipsometer. The technique has been applied successfully not only for the investigation of metal gas reactions but also for studies of oxide film formation on electrodes in electrochemical cells; even if at some wavelengths the liquid is opaque, transparency in some other part of the spectrum allows ellipsometric analysis [3.13]. The original method determines the change of the relative amplitudes and phase factor of the two vector components of a polarized light beam after reflection. As long as only these two ellipsometric measurable quantities at one wavelength are determined as a function of time the quantitative analysis of results has to be made under restricting assumptions. For example a homogeneous, isotropic, nonabsorbing, smooth surface layer must be assumed for a sample. If in addition the angle and the wavelength of the light beam are varied, the experimental data provide much more information and relatively complex structures can be investigated and controlled during processing [3.11, 12]. Modern instruments are equipped with real time data acquisition software and scanning techniques can be used with a lateral resolution down to a few microns. They are available for measurements in air or as in-situ systems for applications under high vacuum conditions and in electrochemical cells. The technique can be used to investigate film samples at various temperatures and pressures and can be combined with other surface analytical methods in the same experimental setup. Quantitative studies can be performed on the oxidation of samples with rough surfaces [3.14] or on the formation of alloy or multilayer structures [3.15] and meanwhile results have been published on many material systems [3.16].

Spectroscopic ellipsometry has evolved rapidly in recent years in both, technology and new applications. Spectral ranges on commercial instruments cover 190 to 14,000 nm, and microstructural and optical analysis now include determination of:
– void or constituent or alloy fraction in a multiconstituent film,
– optically anisotropic materials with arbitraily oriented optic axes,
– multilayer thickness analysis,
– multisample or multidata type (e.g. ellipsometry plus transmittance),
– optical constants and dispersion relations over wide spectral ranges,

– real-time analysis of thicknesses and optical constants,
– surface and interface roughness analysis.

One limitation of the use of ellipsometry is its dependence on the structure and correlation of parameters and variables assumed in models for data evaluation. Measurements of the reflectance and transmittance have the same limitation, but to a greater extent. This limits the complexity of the model. The cross-correlation parameter of the regression fit indicates when the process result may not be unique.

3.5 Energetic Ion Scattering

Another type of experiments developing in recent years for quantitative in-situ measurements of the composition of thin surface layers with a thickness of 1 to 1000 nm is the analysis of the particles emitted from a substrate which is exposed to a beam of high energy ions. The incident beam can consist of H, D, He ions, but for certain applications also heavier ions such as Ne, Ar or even Au are used. Their energies typically range from 1 MeV to several 10 or even more than 100 MeV. If the energy of the ion beam is kept constant, depth profiles of the composition of the near surface region of the target can be derived from the energy distribution of the emitted particles. During the last twenty years progress has been made in the sensitivity and depth resolution of the detecting systems. It is now possible to detect hydrogen and oxygen concentrations as low as 1 at.ppm and to measure concentration changes with a sensitivity of about 1% at high and of 10% at low concentrations, limited primarily by counting statistics. With standard Si-surface barrier detectors depth resolutions in the 10 to 100 nm range are achieved which was extended to the few monolayer range quite recently by using special detection devices. The times required for the measurement of one energy spectrum range from one minute to one hour, depending on the concentration and technique of detection. Thus, ion beam experiments are well suited also for the investigation of the kinetics of metal-gas reactions. Since an overview on this specific aspect of the methods is not available, their performance shall be emphasized briefly in this section.

Radiation damage by the incident ion beam is, except with high energy heavy ions, a less serious problem than one might expect. Firstly, for energetic particles the maximum of damage is at the end of the particle penetration range, i.e., far away from the analyzed near surface region. Secondly, the total defect concentration is still small as well as the amount of sputtered atoms at the high ion energies. Only very high-energetic heavy ions, such as 100 MeV Au ions, produce considerable damage due to the hight amount of energy deposited in a small volume of the crystal .

A big advantage of ion beam experiments when compared to the volumetric method and measurements with the quartz crystal microbalance is the fact that they are not restricted to film samples but any kind of target can be used at any temperature. If gas exposure experiments at pressures higher than ca 10^{-5} mbar are performed the exposure has to be interrupted for each measurement and the pressure reduced to $p < 10^{-5}$ mbar for the time of the measurement.

The basic principles of the techniques, details which are of specific interest for studies on surface layer growth kinetics [3.17, 18] and for the analysis of oxygen [3.19] are described in handbooks. They rely, with the exception of the nuclear resonance technique (Sect. 3.5.3), on the analysis of the energy of the particles leaving the sample. E_0 being the energy of the primary ion beam, the energy E_1 of a particle leaving a collision with a target atom amounts to

$$E_1 = kE_0. \tag{3.4}$$

k is called kinematic factor and in the case of RBS and ERD (Sects. 3.5.1, 2) it depends only on the emission angle of the particle which is counted from the forward direction and the ratio of the primary and secondary ion mass m_1/m_2. Due to the latter relation an identification of the atoms in the sample is possible by analyzing the particle energy. If the collision takes place at depth l in the sample, then the primary ion as well as the secondary ion lose energy due to interactions with target electrons on their way through the target. Therefore, they have lower energies than particles which arise from the sample surface. For shallow depths l the energy loss ΔE is proportional to l and given by

$$\Delta E = \left(k \cdot \frac{dE/dl_1}{\sin \alpha_1} + \frac{dE/dl_2}{\sin \alpha_2} \right) \cdot l, \tag{3.5}$$

where α_1 and α_2 are the entrance and exit angles to the sample surface and dE/dl_1 and dE/dl_2 the energy loss per unit length of the primary and secondary ion, respectively. For larger depths more elaborate formulae have to be used. It is this energy loss which provides the depth scale for depth profiling. Due to the dependence on k it is different for different atom species.

A major problem of all ion beam based techniques is the particle background which arises from the interaction of the ion beam with atoms other than the desired species, for example, with matrix atoms or impurities. Various techniques have been developed to reduce or eliminate this background. A rough discrimination is often possible by using different emission energies E_1 due to differences in the kinematic factor k.

3.5.1 Rutherford Backscattering Spectroscopy

Rutherford Backscattering Spectroscopy (RBS) is primarily used for the analysis of atoms with heavy and medium masses in a lighter matrix [3.18, 20]. Nevertheless, it may also be used for the analysis of oxygen and similar atoms under certain circumstances. Typical primary ions for RBS are He ions with energies in the 1 to 3 MeV range. In RBS the energy $E_1 = kE_0$ of the backscattered He ions is analyzed. For scattering in the exact backward direction the kinematic factor reads

$$k = \left(\frac{1 - m_1/m_2}{1 + m_1/m_2} \right)^2. \tag{3.6}$$

k is always smaller than one and close to one for backscattering from heavy atoms and small for backscattering from light materials. It is this mass dependence which allows us to identify different atom species in a sample. Backscattering on hydrogen is impossible due to kinematic reasons. It is also obvious that particles backscattered by oxygen leave the sample with lower energy than ions backscattered by most metals. Hence, in analyzing the oxidation of metals by RBS one always encounters a background of particles backscattered by the matrix atoms. The Rutherford scattering cross section increases quadratically with the nuclear charge Z of the atoms involved. Therefore, the particle background is in most cases by far larger than the signal from the oxygen atoms and, hence, difficult to subtract. Thus, RBS can be used only for the analysis of the oxidation of light materials such as Mg, Al or lighter ones and only at high oxygen concentrations. Nevertheless, the technique is able to yield absolute concentration values within 1% with reasonable depth resolution down to ca 10 nm. The bulk material serves as reference. With high resolution spectrometers depth resolutions in the few monolayer range close to the surface are achieved [3.21].

At higher energies the scattering cross section of the He ions may exhibit strong deviations from the $1/E^2$-law of the Rutherford cross section σ_R due to the interaction with the atom nuclei. This is most relevant in collisions with light atoms and can even result in pronounced resonances at a resonce energy E_R. At these resonances the scattering cross section can easily be by one order of magnitude higher than s_R and, hence, provide an increased detection sensitivity for the atomic species under consideration. In these cases a peak is visible in the RBS spectrum at the position where the ion energy at the moment of the collision matches the resonance energy. When the energy of the primary ions is increased, the resonance is shifted towards larger penetration depths (accompanied by a broadening of the peak due to energy straggling of the primary beam), thus monitoring the concentration of the atom species under consideration in the sample bulk. A famous example is the resonance of ^{16}O at the ^4He ion energy of 3.045 MeV which is frequently used in the analysis of oxygen [3.22]. Nevertheless the problem of background scattering by the matrix atoms still remains and evaluation of the scattering data is rather laborious.

3.5.2 Elastic Recoil Detection Analysis

Elastic Recoil Detection Analysis (ERDA) is the ideal method for the analysis of hydrogen and other light atoms such as oxygen and others [3.18]. A beam of heavy primary ions is required, starting at He as the lightest, for the analysis of hydrogen, continuing with Ne and Ar, and going up to really heavy ions like Au or J. The energies required range from a few MeV for He up to more than 100 MeV, for example, with Au. In ERDA target atoms are ejected – so-called recoil particles – in single collisions by the ions of the primary beam and their energies analyzed. The recoil energy E_1 amounts to $E_1 = kE_0$ with the kinematic factor in this case given by

$$k = \frac{4m_1/m_2}{(1 + m_1/m_2)^2} \cos^2 \vartheta. \tag{3.7}$$

k is small for low as well as for high mass ratios and reaches a maximum for equal masses. It is 0 for $\vartheta = 90°$ and has its maximum for $\vartheta = 0°$ (foreward scattering). Furthermore, the emission of recoil particles is only possible up to a maximum angle smaller than 90°. Therefore recoil particles have to be detected in the forward direction with angles of the order of 40°. The ERDA scattering cross section increases, due to kinematic reasons, with approximately the 4th power of the nuclear charge of the primary ions. Therefore, the use of heavy primary ions is advantageous. On the other hand, it is almost independent on the nuclear charge of the target atoms. Mass and depth information is obtained in a similar way as by RBS.

A more serious problem in the application of ERDA is discrimination of the ERD-particles from the normally high background of primary ions backscattered by the matrix atoms. In the case of hydrogen this is commonly done by the use of a thin shielding foil in front of the detector. The thickness of the foil can be chosen such as to stop the backscattered particles completely, but to allow the recoil hydrogen ions to penetrate with only minor energy loss. This technique is successfully applied to the analysis of very light atoms such as H, D, He, and Li to a total depths around 1 mm and with depth resolutions in the 100 nm range. The depth resolution is limited by the energy straggling in the detector foil. With special precaution sensitivities around 1 at.ppm can be achieved. Absolute contentration determination of hydrogen is possible within about 1% if a reference sample is available. Otherwise, the errors are in the 10% range, primarily due to uncertain values for the specific energy loss of the ions.

Unfortunately, the foil technique cannot be applied to the analysis of atoms such as oxygen, since the ion mass ranges of oxygen and the primary ions in question are of about the same size. But in contrast to RBS, where the backscattered particles are indistinguishable He ions , only with different energy , the ERD-particles bear their element information within themselves. Thus, by identifying the recoil particle and simultaneously measuring its energy, one is able to measure almost background free depth profiles. For this purpose several concepts have been developed. The more widely used ones are time of flight telescopes in combination with standard Si-surface barrier detectors and multiwire chambers. In both cases high energy heavy ion beams have to be used (e.g., 40 MeV Cl ions). This technique is limited to high energy tandem accelerators. The depth resolutions lie in the range of 10 to 50 nm because of the finite energy resolution of the detectors, unless high resolution spectrometers for heavy ions are used [3.23] which are in general bulky and not commonly available. The high specific energy loss per unit length of heavy ions is favorable for high depth resolution, but it may cause serious radiation damage since high amounts of energy are deposited in a small volume of the sample. Almost background-free depth profiles of oxygen could be obtained in ERD-experiments using heavy ions (Ne, Ar) in the few MeV range only recently by combining a high resolution electrostatic spectrometer with a time of flight technique [3.24, 25]. By this method depth resolutions in the few monolayer range were attained. Similar depth resolutions in the analysis of hydrogen have been achieved by the same group [3.26].

3.5.3 Nuclear Reaction Analysis

Nuclear Reaction Analysis (NRA) uses nuclear reactions between the incident ion beam and the atom species to be analyzed for depth profiling purposes [3.17, 18]. With respect to oxidation and hydrogen absorption of metals, two nuclear reactions must be discussed in more detail: The reaction $^{16}O(d,p)^{17}O$ which is frequently used for the analysis of oxygen and the resonance reaction $H(^{15}N_{\alpha,\gamma})^{12}C$ which is the standard reaction for high precision hydrogen depth profiling.

The first process belongs to a group of reactions which require ion beams of light ions such as H, D, 3He with energies in the few MeV range. For the $^{16}O(d,p)^{17}O$ reaction a deuteron energy of 0.9 MeV is most appropriate, since the reaction cross section exhibits an almost flat part in the energy interval from 0.9 to 0.7 MeV. This is important, since most nuclear reactions have cross sections with various resonances which make the evaluation of experimental data difficult or impossible. The energy $E_1 = kE_0$ of the emitted protons amounts to 1.726 MeV for an emission angle of 164°. In the case of nuclear reactions k is a function which depends in a complicated way on the masses of the involved particles, the emission angle, the released energy of the reaction and the energy of the primary beam [3.17]. Thus, the notation $E_1 = kE_0$ is just formal. The identification of different atom species can be done by a proper choice of the nuclear reaction, of the primary ion beam and – to a certain extent – by the analysis of E_1 which is different for different reactions. Interfering reactions must be avoided whenever possible. Depth information is obtained similarly to RBS and ERDA.

A problem which is as serious as in ERDA is the reduction of particle background, particularly the large amount of primary ions backscattered by the matrix atoms. In order to avoid this part of the particle flux, commonly thin shielding foils are used in front of the detector which allow the reaction products to penetrate, but stop the primary ions. In case of the reaction $^{16}O(d,p)^{17}O$ a 10 mm thick Al foil is sufficient. This foil, however, rules out reactions with low energy release and with heavier reaction products, such as α-particles and, in particular spoils the energy resolution of the detector set-up. Thus, typical depth resolutions lie in the 100 nm range with total available depths of several micrometer. Another problem is background signals from parasitic reactions with other light atoms present in the sample. This is particularly important when profile measurements at low concentrations are performed or when deuteron or ^3He ion beams are used. This background limits common NRA reactions to a minimum concentration of the order of 100 at.ppm. However, it can be reduced by about two orders of magnitude by a technique, first proposed by *Carstanjen* et.al [3.27]. It is based on the fact that in many nuclear reactions the remaining nucleus, in our example the ^{17}O nucleus, remains in an excited state which decays by the emission of gamma-rays. Counting the reaction products in coincidence with the emitted gamma-rays reduces the background signals caused by parasitic reactions considerably. By this technique oxygen concentrations down to 1 at.ppm should be detectable.

The second reaction mentioned, $H(^{15}N_{\alpha,\gamma})^{12}C$, for the analysis of hydrogen belongs to the group of resonance reactions. These reactions exhibit a resonance of width Γ in the reaction cross section at the resonance energy E_R of the primary particles. The resonance energy of the ^{15}N-ion beam is 6.394 MeV and the width is 1.85 keV [3.28]. Depth profiling is done in these cases by increasing the energy of the ion beam, starting at the resonance energy. Thus, the resonance is pushed into the sample and the reaction yield, in the present example the gamma-yield, is measured in dependence on the energy of the ion beam. This is often done by applying a bias voltage to the target, but also more elaborate devices are available. By these techniques the evaluation of data, for example, the determination of a depth profile, is rather simple. If the resonace width is narrow, excellent depth resolutions are achieved. With the reaction $H(^{15}N_{\alpha,\gamma})^{12}C$ depth resolutions of 3 nm, at oblique incidence also of 1 nm have been obtained in the analysis of hydrogen in Si. In fact, the resonance is that sharp that the depth resolution is almost completely dominated by the Doppler-shift due to the motion of the hydrogen atoms. If special precautions are met, even very small hydrogen concentrations can be monitored which are of the order of 1 at.ppm. Unfortunately, most nuclear reactions have low cross sections when compared with RBS and ERD, typically less than 1 barn. Therefore, commonly large detectors are required for particle or gamma-ray acquisition.

It should be mentioned that a variety of nuclear reactions with rare isotopes of light elements, such as ^{13}C, ^{15}N or ^{18}O is available. The $H(^{15}N_{\alpha,\gamma})^{12}C$ reaction for the analysis of hydrogen is such an inversed reaction with ^{15}N. These reactions can be used for the measurement of tracer diffusion in the presence of the abundant stable isotopes. This is of particular interest for the very important question in metal oxidation, whether the metal or the oxygen atoms are the diffusing species.

3.6 X-Ray Reflectivity

The thickness and structure of thin surface films can be determined accurately by X-ray reflectivity [3.29,30]. This nondestructive and quantitative technique is used widely for the characterization of multilayers and also well suited to investigate oxidation processes due to its nanometer scale sensitivity and the possibility to characterize a sample in situ prior to an oxidation experiment with respect to surface roughness and the presence of other films. In a standard reflectivity experiment a monochromatic X-ray beam is incident under grazing angles on a liquid or solid sample. The specular reflected intensity is measured over several orders of magnitude as a function of the angle of incidence, α, which ranges typically from $0.1°–2°$.

X-ray reflectivity can be described in terms of classical optics as Fresnel reflectivity. For X-rays the appropriate index of refraction is $n = 1 - \delta$ with

$$\delta = \frac{r_0}{2\pi}\lambda^2\rho_e, \tag{3.8}$$

where r_0 denotes the classical electron radius, λ is the X-ray wavelength, and ρ_e the electron density of the material. Corrections that consider dispersion and absorption are discussed in ref. [3.31]. For typical X-ray wavelengths and most materials δ is of the order 10^{-6}. The refractive index is less than one and, therefore, matter is optically less dense than vacuum or air and an X-ray beam is totally reflected at a sample surface for incident angles below a critical angle, $\alpha_c = \sqrt{2\delta}$. For higher angles the reflectivity drops quickly with α^{-4}.

The information obtained by specular reflectivity experiments is the average electron density profile perpendicular to the sample surface. For a simple homogeneous sample, the critical angle gives the electron density or the mass density if the chemical composition is known. Surface roughnesses below 10 nm are determined from deviations from the ideal Fresnel behavior.

The thickness of an oxide layer is detected by oscillations in the reflectivity if the electron density differs from the underlying material. In practice, the thicknesses of oxide layers ranging from 1–200 nm can be determined with a typical error of 1%. Thicker layers usually cause resolution problems due to very narrow oscillations, and the extended oscillations of monolayer films are difficult to observe for intensity reasons. The electron density of the oxide film is obtained directly from the critical angle if the film thickness exceeds the X-ray penetration depth of several nanometers in the total reflection region. Otherwise, the density can be derived from the oscillation amplitude. Usually all parameters, including surface and interface roughness, are determined by simulation programs. Typical error ranges are 1% for the density and 0.1 nm for root mean square roughnesses if absorption, dispersion, and multiple scattering are taken into account. Recent studies of thin film oxidation where given in [3.32,33].

X-ray reflectivity can also be measured by energy-dispersive methods where a continuous X-ray spectrum is used at fixed scattering angles. As the time of several hours required for a measurement with X-ray tubes is reduced to 15 minutes, this variant is especially suited to investigate the kinetics of low-temperature oxidation. A recent example is the determination of the time law of liquid Ga oxidation [3.34].

The use of X-ray reflectivity offers several advantages: it can be applied both, to amorphous and crystalline oxides, no sample preparation is required, and bulk samples as well as thin films on a substrate at any temperature can be used. The electron density and thickness of thin films are determined independently since the refractive index is very close to one. On the other hand, grazing angle experiments require precise sample alignment. Grazing angles also cause a sample illumination which is prohibitive for most spatially resolved investigations. Reflectivity data from curved samples must be interpreted with care due to finite resolution effects.

3.7 Surface-Analytical Methods

For the investigation of solid surfaces, powerful techniques have been developed in the past decades [3.35] and a large stock of experimental and theoretical data meanwhile has been compiled [3.36]. A beam of electrons, ions, or photons is directed to the surface of a substrate and the quantity and/or energy of the electrons, ions or photons emitted is analyzed. More than 60 different experimental methods can be distinguished if all variants are considered. The atomistic structure of surfaces is studied by low energy electron diffraction (LEED) [3.37] and, in recent years, by scanning tunneling spectroscopy (STM) [3.38–40]. The electronic structure and chemical bonds of surface atoms are determined by ultraviolet photoemission spectroscopy (UPS) or X-ray photoemission spectroscopy (XPS) [3.41] and the chemical composition of surfaces can be analyzed by secondary ion mass spectroscopy (SIMS) or Auger electron spectroscopy (AES). Other techniques investigate electronic, magnetic, and vibrational properties of surfaces and adsorbed species. The interest of surface science is presently focused mainly on the characterization of surfaces of pure metals and semiconductors and on catalytic reactions of gas molecules on various surfaces [3.36]. The number of papers published on properties of oxide surfaces and species adsorbed on oxides is limited when compared with the activity in other fields.

With the big arsenal of sophisticated and sensitive methods of surface science, the interactions of oxygen and of other gases with metal surfaces can be investigated in great detail up to the range of few monolayers and valuable information on the character of bond states, the structure and chemical composition of surfaces is obtained. The signals measured are mainly caused by the surface atoms. Contributions from subsurface atoms decrease rapidly with the distance from the surface. Surface science techniques are, therefore, not well suited for quantitative determination of the growth rate of oxide scales thicker than a few monolayers. The methods suffer also from another hindrance. They have to be perfomed under ultra-high-vacuum conditions. In most papers the property changes caused by gas absorption are plotted as change of an electronic signal versus the dose of impinging gas molecules. The *dose*, also called *exposure,* is equivalent to the product pressure times time. It is a qualitative measure and replaces the not exactly known parameter layer thickness in the discussion of results. However, the reaction probability of gas molecules can be reduced in not an exactly predictable way by orders of magnitude after the surface has been covered by more than one monolayer. Therefore, increasing exposure in a plot means only *more absorption*, but frequently one cannot decide whether it is much more or only a little bit more.

A reliable surface analytical method for the investigation of an oxidation process would be checking the thickness of a set of surface layers after different oxygen exposures by in-situ depth profiling, for example, in combination with AES. The thickness of the oxide scale is then determined by the sputter time needed for the removal of the oxide scale by ion bombardment. If only the final thickness of an oxide scale after extended exposure must be measured, ex-situ techniques can be applied which are capable to determine thin surface layers, for example scanning or transmission electron microscopy (SEM,TEM) [3.42, 43]. In this case it is assumed that the layer thickness is not changed anymore by an air exposure.

A field where surface analytical methods can contribute valuable information to the understanding of low-temperature oxidation kinetics would be studies on the behavior of surface charges on oxide surfaces in the early stage of oxidation. The conditions necessary for the establishment of the Mott potential across a growing oxide layer are quite different for different metals. Only little experimental evidence is available on absolute values of surface excess charges on oxides and their dependence on layer thickness and the gas pressure. An interesting experiment would be to sustain the Mott potential across the oxide layer by an electron-beam bombardment of the oxide surface beyond the thickness where the electronic current through the oxide breaks down. Under this condition the high oxide growth rate found in stage one of the models (Chap.5) should be maintained and thicker oxide scales should be formed. Results of this type would provide experimental evidence of the close interrelation between the Mott potential and the thickness of natural oxide scales on metal surfaces at low temperature.

4 Hydrogen Reactions

In metal-gas systems two absorption processes can be distinguished, solution of atoms from gas molecules in liquid or solid metals and formation of compound layers on the metal surface. At temperatures below 300 °C the interactions of gases with the bulk of a metal phase are normally sluggish. Only hydrogen atoms can move fast enough in a metal matrix to yield perceptible concentration changes within short times by diffusion controlled processes. The surface of metal samples is normally covered by the natural oxide skin and, therefore, studies on reaction kinetics in metal hydrogen systems under clean conditions are troublesome. Despite this handicap, it is very informative to investigate the reaction mechanisms of ideal metal hydrogen systems because they are prototypes of basic processes which also occur in more complex reactions of other metal-gas systems. The concentration changes caused by reactions of nitrogen, hydrocarbons or ammonia are, for example, restricted to a narrow area close to the metal surface at low temperatures. But the growth of such thin surface layers and of thicker, adherent nitride or carbide coatings at higher temperatures can be described by the same mechanisms as presented in this chapter for hydride formation if no semiconducting phases are involved. In semiconducting layers of oxides or chlorides electric currents of ions and electrons have to be considered and effects of electric fields created by charged particles. Thus, terms for the local electric potential must be inserted in the transport equations and the mathematical treatment becomes more difficult.

The reactions of hydrogen gas molecules with a pure metal phase are the most simple chemical processes conceivable for a metal-gas system. Reactions of gas molecules consisting of different atoms, for example, H_2O or CO, with metallic alloy phases are not much different in principle but the volume of the mathematical problem is increased remarkably. The content of this chapter will be confined to the discussion of simple, but still complete, reactions in pure metal-hydrogen systems. This may avoid confusion and improve the understanding of the role of individual partial processes and their interrelations within a model better than a more detailed description of the real reactions responsible for the behavior of specific metals in hydrogen gas under less clean conditions.

In Sec. 4.1 some basic properties of metal-hydrogen systems and typical results of experimental investigations are presented. The characteristic features of reaction models are demonstrated in Sec. 4.2 by rate and time laws derived for the limits where only one of the partial steps is rate determining. In Sec. 4.3 computer simulations for an advanced and more realistic model are discussed. They demonstrate the power of reaction models with only one rate determining step, but also their inherent limitations and show how the kinetics of models with more than one rate determining step look like.

4.1 Experimental Results

4.1.1 Metal-Hydrogen Systems

A metal-hydrogen system is called an exothermal system when the heat of solution or of the hydride formation is negative, which means energy is evolved to the surrounding by the reaction. If heat energy is consumed during the solution process, the reaction is endothermal and normally no hydrides are formed. Exothermal systems are the hydrogen systems of the alkali and earth-alkali metals and the transition metals of group III-B to group V B of the periodic table of elements. The other metals can be attributed to the group of endothermal systems with palladium as an exception [4.1–8].

Endothermal systems show a limited solubility for hydrogen in the ppm level which is increased at constant pressure with increasing temperature. Dissolved hydrogen is normally quenched in after melting or annealing processes. Despite the small hydrogen concentration severe material problems can arise in these systems since the equilibrium pressure of H_2 gas increases enormously at lower temperatures. This can, for instance, cause bubble formation in aluminum or hydrogen embrittlement in steel if hydrogen is trapped at dislocations or crack tips [4.9, 10].

Exothermal systems are capable to dissolve large amounts of hydrogen or to form hydrides. The Figs. 4.1, 2 show temperature vs. concentration plots of the systems Ta–H and Hf–H with H_2 isobars. In the Ta–H system the solubility of hydrogen in the metal matrix is very high and at temperatures above 50 °C no hydrides are formed. In the Hf–H system the α metal phase can dissolve only 1 to 10 at. %H, depending on the temperature. At higher hydrogen concentrations δ HfH_2 hydride is formed. The horizontal isobars in the two-phase region of Fig. 4.2 represent the decomposition pressure or plateau pressure of the hydride. At constant temperature hydride is formed only if a H_2 pressure higher than this value is applied. This behavior of hydride formers to absorb large amounts of hydrogen gas if the pressure is increased above the plateau pressure and to release the gas

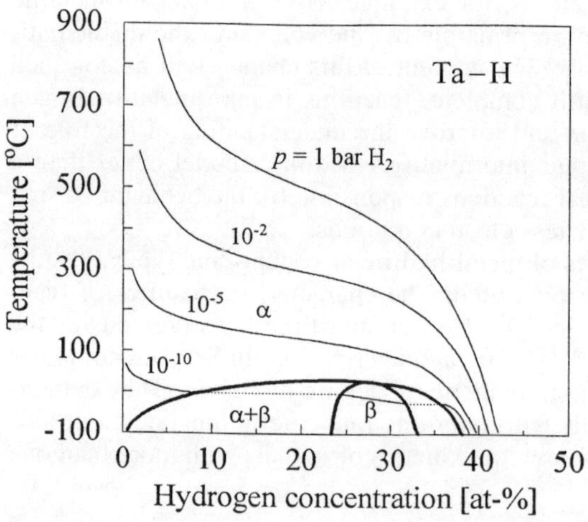

Fig. 4.1 System Ta–H in a presentation temperature vs. concentration with H_2 isobars[4.8]

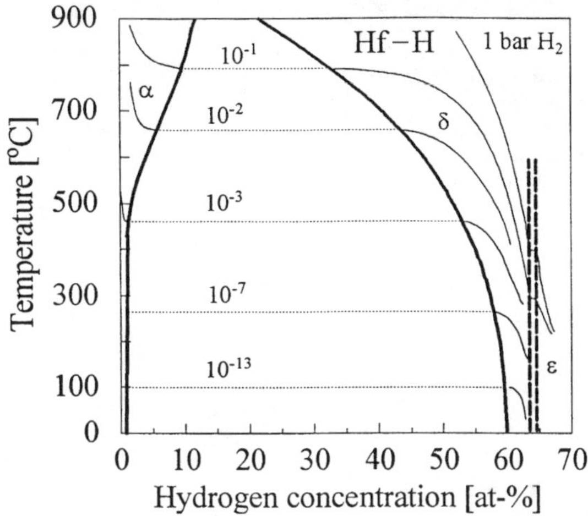

Fig. 4.2 System Hf–H in a presentation temperature vs. concentration with H_2 isobars[4.8]

again if it is lowered below this value is a basic requirement for their use as hydrogen storage materials. Since heat energy is evolved as reaction enthalpy during hydride formation and absorbed during decomposition, metal hydrides are also ideal working substances for cyclic loading processes in heat pumps and heat exchangers [4.11].

The kinetics of hydrogen absorption and desorption reactions at ambient and low temperatures is a very fast solid state reaction because of the high diffusion rates of hydrogen in metals and metal-like hydrides. The diffusion coefficients are in the order of magnitude of 10^{-8} to 10^{-4} cm^2/s. Equilibrium of a diffusion controlled reaction is established in samples with a thickness of about one mm after only ten minutes to few hours and for powder particles of 10 μm diameter after about one second or less. Since the mobility of hydrogen atoms in the bulk of metals or in hydride phases is high, sluggish reactions observed with thin samples such as films, powders, or surface coatings must be caused by slow reaction rates of the partial steps in the surface region. Slow dissociation of chemisorbed H_2 molecules on the sample surface or slow permeation of hydrogen atoms through a contamination layer are processes which can impede or sometimes even prevent a hydriding reaction at low and slightly elevated temperatures. Strong reduction of the hydrogen absorption kinetics by surface poisoning can also have beneficial aspects. Hydrogen uptake from the moisture of the atmosphere is avoided in materials which are susceptible to hydrogen embrittlement or bubble formation.

The reaction mechanism of hydride formation is schematically shown in Fig. 4.3a. If an adherent hydride scale is growing on a metal sample the concentrations of hydrogen in the physisorbed and in the chemisorbed state can be quite different. The concentration in the first subsurface layer in the β hydride phase has a value between the equilibrium concentration given by the H_2 pressure in the gas phase for the temperature chosen (isotherms in Fig. 4.2) and the plateau pressure. At the β/α interface the concentration in both phases is defined by the two-phase equilibrium condition and determined by the end points of the plateau-pressure line in

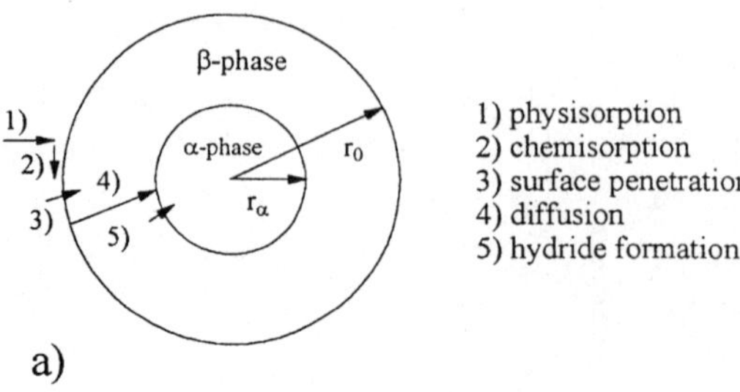

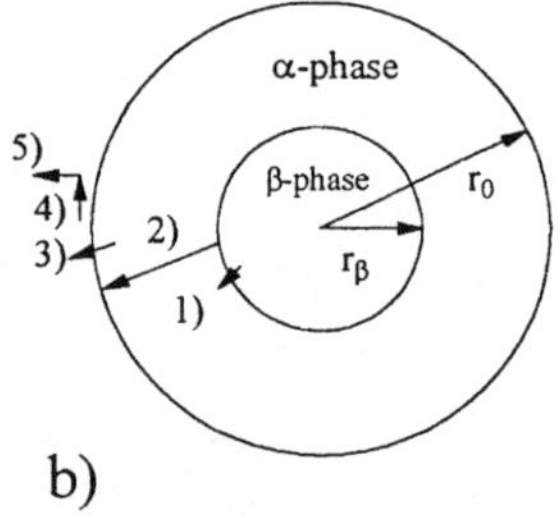

Fig. 4.3 Mechanism of (a) hydride scale formation and (b) hydride decomposition, schematic

Fig. 4.2. If the mobility of hydrogen in the hydride is high, the surface concentration attains a value close to the α/β equilibrium concentration and the concentration gradient ∇c in the hydride scale is small. Low mobility in the β phase raises the hydrogen concentration in the subsurface layer to the equilibrium value given by the H_2 pressure in the gas phase. Whether or not a larger concentration gradient exists inside the α core of a sample depends on the relative mobility when compared with the rate of other partial steps of the overall reaction. Normally this partial step is fast and saturation of the α phase is established within very short time.

Decomposition of a hydride particle is accomplished by a different mechanism. Here a surface layer of the α phase is formed and the hydrogen atoms are produced by decomposition of the β hydride at the α/β interface and diffuse through the metal layer to the surface of the sample (Fig. 4.3 b). When the H_2 pressure in the reactor is very low, the concentration of physisorbed hydrogen molecules is very small. The concentration of chemisorbed hydrogen species is small when the recombination of chemisorbed atoms to physisorbed molecules is fast. The hydrogen concentration at the subsurface layer of the α phase is only small, if in addition the interface penetration reaction is also a fast process, then the subsurface region is in equilibrium with the gas phase. The hydrogen concentration in the α phase at the α/β interphase attains the terminal solubility value when the diffusion of hydrogen atoms through the α phase is the slowest step of the reaction. For a mechanism where desorption of chemisorbed hydrogen atoms is rate controlling this, concentration value for the two-phase equilibrium is established everywhere inside the metal surface layer.

The kinetics of an absorption experiment cannot always be described correctly by models derived for the limit of only one rate determining step. Whether or not this approximation can be used is indicated by the pressure dependence of experimental rate and time laws. This will be demonstrated in subsequent sections. In isothermal experiments the absorption rate is given by a relation v^n with $0.5 < n < 1$. If exponents in between the limits 0.5 and 1 are observed a mechanism with more than one rate determining step is active.

Experimental results measured with real materials can also be affected strongly by some effects not mentioned so far. In addition to surface poisoning cracks can be formed in surface layers due to the expansion and shrinking of a sample during hydriding and dehydriding. This can cause a reduction of the size of powder particles and spalling of contamination layers. The extended incubation periods frequently observed in absorption experiments are mainly caused by these processes. The increase of the reaction rate after the initial slow region can be attributed to the creation of unpoisoned reactive surface areas. The discussion of this kind of technical kinetics observed during handling and activation of hydride materials under real conditions of practical application is beyond the scope of this book and the reader is referred to the relevant literature, for example, *Gérard, Ono* ref. [4.5, pp.165–195].

4.1.2 Hydrogen Solution in Metals

Isothermal isobaric experiments
The experimental investigations on hydrogen absorption kinetics can be subdivided into various types. The standard methods are exposure experiments performed under isothermal isobaric conditions where the amount of absorbed hydrogen gas is measured as a function of time, for example, by the weight gain of a sample. If the change of the electric conductivity of a wire sample is proportional to the hydrogen concentration, the time law can be determined by curves resistivity versus time. The average concentration of the sample is then given by the resistivity change. For reactions where the rate controlling partial step is on the surface no problems arise since the concentration in the sample is constant anyway. Figure 4.4 shows results of this kind of experiments for the system Ta–H [4.12]. The curves begin with a constant initial absorption rate which increases with increasing temperature. After some minutes the concentration approaches the equilibrium value which is given by the phase diagram shown in Fig. 4.1. Desorption curves measured with the same technique are dipicted in Fig. 4.5 [4.13]. Also in these experiments the constant initial reaction rate increases with temperature.

Manometric and volumetric techniques
The hydrogen absorption kinetics of materials available only as powder or granules can be measured by so-called concentration-pressure isotherms. The sample is given to a closed reactor of known volume. Then, hydrogen is added and the pressure rises to a high initial value. The amount of hydrogen absorbed during exposure can be calculated from the pressure reduction if the volume of the gas phase and the mass of the sample are known. Typical results of this type of experiments are presented in the next section.

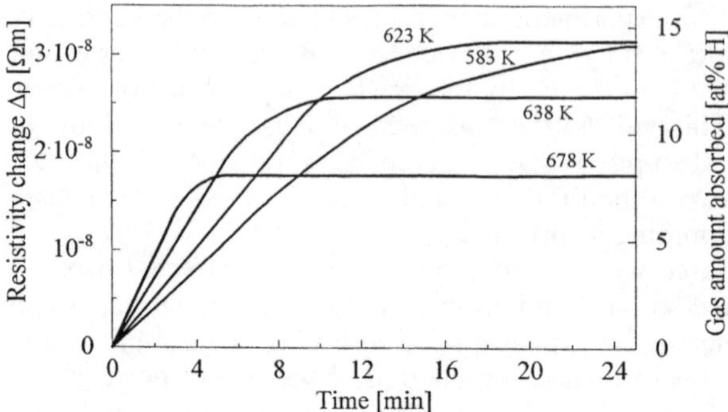

Fig. 4.4 Hydrogen absorption of a tantalum wire at 2.5 mbar H_2 pressure, measured by the change of the electric resistivity [4.12]

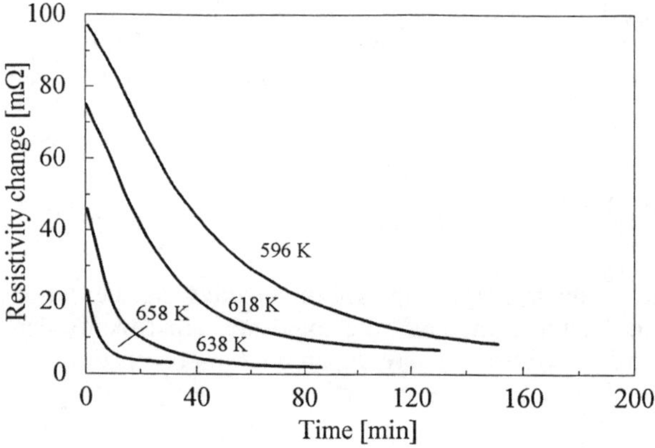

Fig. 4.5 Resistivity change during hydrogen desorption from a niobium wire annealed in high vacuum [4.13]

Instead of admitting a known gas amount at the beginning of an experiment also a known gas flux can be maintained to a closed reactor, for example, through a calibrated capillary (Sect. 3.2). The reaction rate is then given by the flux rate minus the gas amount required to increase the pressure in the reactor. If this reaction rate is compared with the impingement rate of gas molecules, the *sticking coefficient* can be determined immediately. This is the probability that an impinging gas molecule is not reflected but undergoes a reaction. Since part of the atoms present in a gas molecule can also be desorbed again, for example, H_2 from H_2O, the meaning of the term *reaction probability* is more precise. This method has successfully been applied for the investigation of film samples deposited under ultrahigh vacuum conditions [4.14–20] but it is not restricted to this form of samples. Figure 4.6 exhibits results of two exposures of tantalum films to hydrogen at 77 and 298 K [4.21]. The absorption probability of hydrogen molecules begins in both cases at the maximum value one. At room temperature the reaction probability is reduced by more than one order of magnitude within the first monolayer. Tantalum hydride is formed with an almost constant reaction probability until the

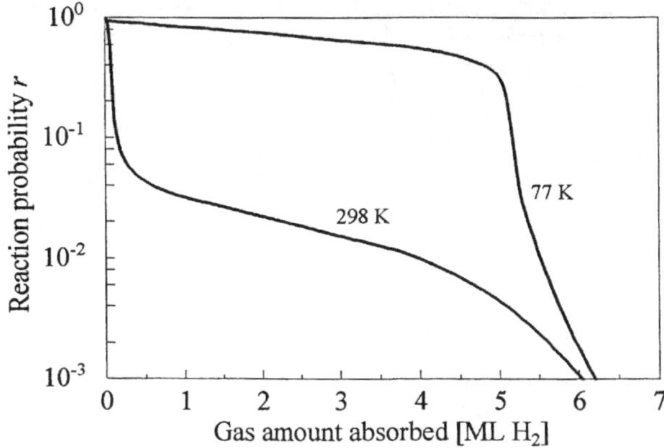

Fig. 4.6 Hydrogen absorption of tantalum films at 77 and 298 K, measured with the volumetric Wagener method [4.21]

film is totally reacted to TaH (Fig. 4.1). At low temperature the surface coverage by the molecular precursor state is much higher and enhances the reaction rate. The reaction probability is close to unity almost up to the end of the run. This finding indicates that dissociation of physisorbed hydrogen molecules on the surface is the rate determining step in this experiment and not diffusion of hydrogen atoms in the bulk of the sample.

4.1.3 Hydride Formation

In binary metal-hydrogen systems hydrides can be formed only if the hydrogen pressure exceeds the decomposition pressure or plateau pressure of the hydride. In Fig 4.7 two pressure-concentration isotherms are shown schematically for two different temperatures. The branch of the curves on the left side represents the pressure dependent equilibrium concentration of the α phase according to Sieverts' law. The horizontal lines are the plateau pressures and on the right side the steep increase of the equilibrium pressure with concentration in the β phase is shown.

Isothermal isobaric experiments
The dashed horizontal line in Fig. 4.7 displays two possible results obtained by this kind of experimental procedure. At the higher temperature $T + \Delta T$ the hydrogen absorption cannot exceed the equilibrium concentration A in the α phase of only a few percent. However, if the temperature is lowered to T, the exposure pressure is higher than the plateau pressure and the sample can absorb hydrogen until it is totally converted to hydride with the equilibrium composition B. The shape of experimental absorption curves plotted in a concentration versus time diagram does not differ much in both cases and looks similar to the curves depicted in Figs. 4.4, 6. After a fast linear initial absorption region the reaction rate is becoming very slow when the equilibrium concentration is approached.

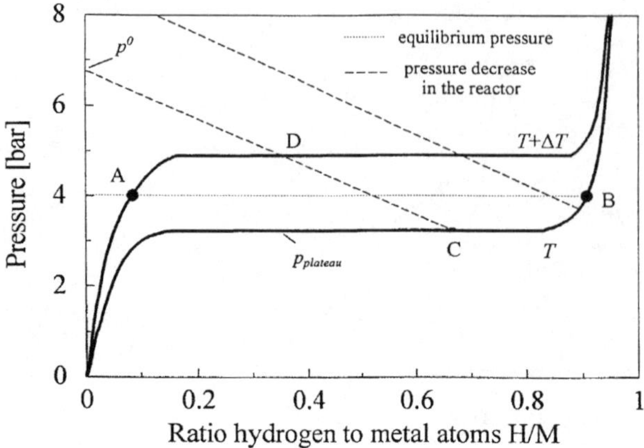

Fig. 4.7 Pressure-concentration isotherms of a hydride forming system, schematic

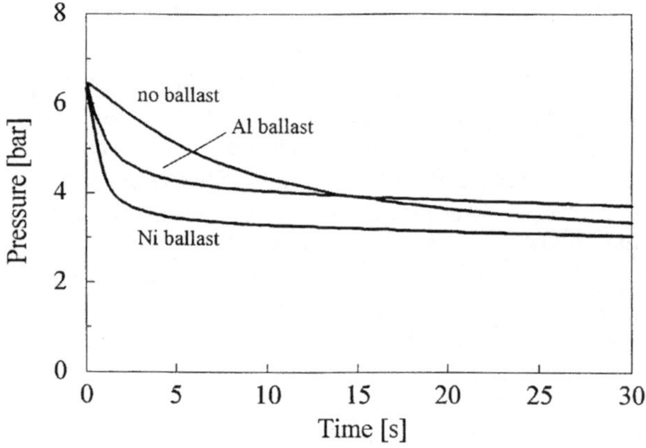

Fig. 4.8 Absorption kinetics of LaNi$_{4.7}$Al$_{0.3}$ powder at 343 K without and with ballast of Ni and Al powder [4.22]

Manometric and volumetric methods

Extensive experimental investigations have been performed with powder samples of alloys which are candidates for storage materials by measuring concentration-pressure isotherms. With this technique the initial exposure pressure in the closed receiver is decreased during the absorption run and the reaction stops when the equilibrium pressure at the temperature chosen has been reached. If not enough hydrogen had been admitted to the reactor only part of the sample is converted to hydride (point C in Fig. 4.7). The heat of reaction, ΔH, produced during hydride formation creates a serious problem in experiments with powder samples. The temperature of the particles is increased and if the decomposition pressure of the hydride is approached the reaction rate is slowed drastically. This is demonstrated by the curves in Fig. 4.8. The reduction of the pressure in the receiver caused by the absorption process represents the reaction rate. It is increased if the temperature change of the powder sample is kept small by adding aluminum or nickel powder as thermal ballast.

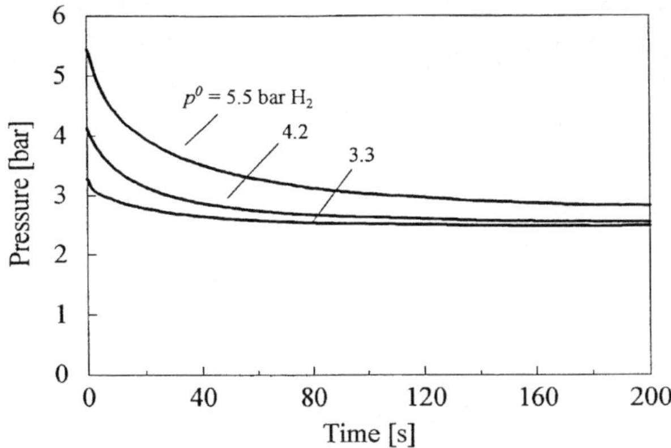

Fig. 4.9 Kinetics of hydrogen absorption of LaNi$_{4.7}$Al$_{0.3}$ powder measured by concentration - pressure isotherms [4.22]

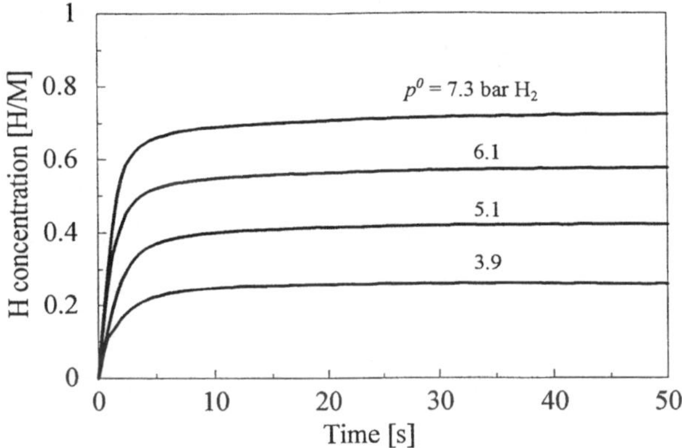

Fig. 4.10 Absorption kinetics of LaNi$_{4.7}$Al$_{0.3}$ powder presented in a plot concentration vs. time for various values of the initial loading pressure [4.23]

Fig. 4.9 presents a typical set of absorption curves for LaNi$_{4.7}$Al$_{0.3}$ powder, an alloy frequently used as hydrogen storage material [4.22]. The absorption rate is high for this material and the powder particles of about 10 µm diameter are hydrided in about 20 seconds. Depending on the initial loading pressure the final concentration and, thus, also the equilibrium pressure of the hydride sample is increased. The results can also be presented in a plot concentration versus time (Fig. 4.10). The initial reaction rate increases at higher initial pressure but the total time for loading is with about 5 s almost constant and very short [4.23].

The kinetics of the degassing process can be measured if the hydrogen evolved from the sample under high vacuum conditions is collected in a closed reservoir of known volume. The pressure increase in the reservoir corresponds to the concentration decrease in the sample. Figure 4.11 shows results of such desorption experiments at 570 K for magnesium powder samples with different initial hydride content. A set of powder samples degassed at different temperatures is depicted in Fig. 4.12 [4.24]. The reaction rate is strongly temperature dependent.

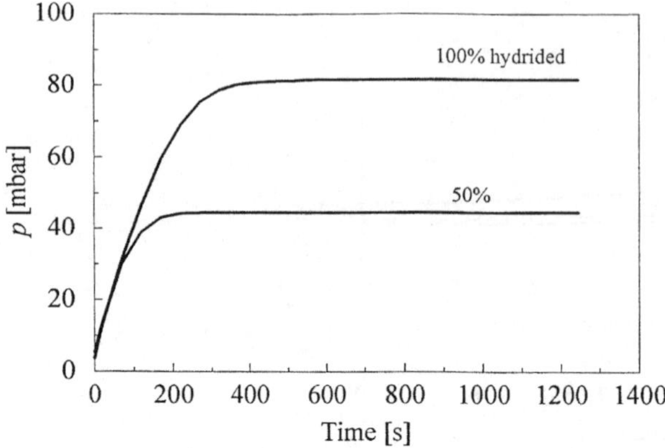

Fig. 4.11 Desorption kinetics of magnesium powder samples with different initial hydride content at 570 K. The amount of desorbed hydrogen H/M is determined from the pressure increase in a large closed receiver

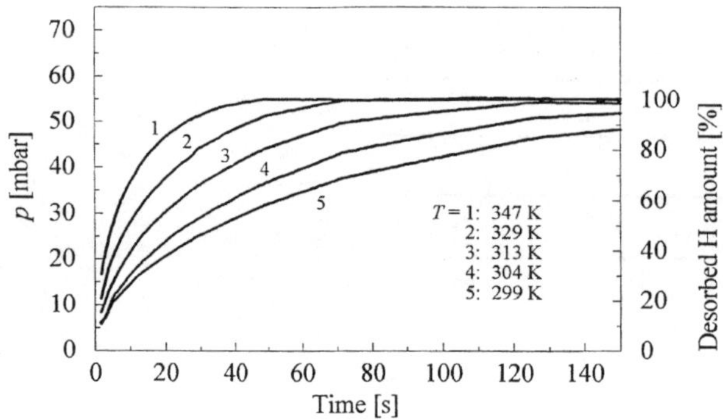

Fig. 4.12 Desorption kinetics of $LaNi_{4.7}Al_{0.3}$ powder samples measured at different temperature. The amount of desorbed hydrogen H/M is determined from the pressure increase in a large closed receiver

Concluding Remarks
At the end of this section it should be emphasized again that the examples presented are results of experiments performed under conditions where the sample surfaces can be considered unpoisoned or at least not severely poisoned. They are not typical for the kinetics observed under less pure conditions. This selection had to be made since the models on kinetics in ideal metal-hydrogen systems discussed in subsequent sections of this chapter are restricted to this situation, too. A poisoned surface is the normal state for a metallic sample investigated under atmospheric or high vacuum conditions. The reaction mechanisms of hydriding and dehydriding processes are frequently changed substantially and the shape of absorption and desorption curves then looks quite different. Commonly oxide layer formation on the sample surface is responsible for poisoning. Reaction mechanisms active under these more realistic conditions are discussed in more detail in Chap. 6. Since some knowledge on oxidation processes at low temperature is desirable for the understanding of poisoning problems, the treatment of this topic has

been shifted behind the chapter on oxidation kinetics. However, one should be aware that still many phenomena observed in the field of reaction kinetics of poisoned systems are not yet well understood.

4.2 Hydrogen Solution in Metals

4.2.1 Reaction Mechanism, Partial Steps

In calculations of rate and time laws for the limit with only one rate determining partial reaction it is assumed that the system is in a state far from the equilibrium state. Far means less than about 50% saturation or a factor of two above or below the equilibrium pressure. If the system approaches the equilibrium state, models based on this approximation have also to consider the backward reaction of the rate determining partial step. Despite the severe restrictions introduced by the simplified reaction mechanism the rate and time laws for such models still can be lengthy mathematical expressions.

Solution of atoms from the gas phase in a condensed phase is one of the most simple metal-gas reactions. The reaction mechanism combines similar processes as discussed in Sect. 2.2.1 where the conception of partial steps as elements of a reaction mechanism have been explained and the procedures for how they must be linked to define a model.

The following four reaction partial steps are needed to describe a solution reaction:

$$H_2 \leftrightarrows 2H \text{ (dissolved in M)}$$

(1) H_2(gas) $\leftrightarrows$ H_2 (physisorbed on M),
(2) H_2(phys) $\leftrightarrows$ 2 H(chemisorbed on M),
(3) H(chem) $\leftrightarrows$ H(dissolved in M),
(4) diffusion of H in M.

The rate laws of the forward and backward partial reactions v_{i+} and v_{i-} are flux densities given as number of particles per area and time, measured for example in units ML/s $= 10^{19}$ particles m^{-2} s^{-1}.

The partial steps 1 to 3 are processes on the sample surface. If one of them is rate determining, the shape of the sample enters the rate or time law only as a factor S/V which is the ratio surface area S to volume V. The concentration c of the sample in particles/m^3 is then obtained by

$$(c - c_0) = \frac{S}{V} \int_0^t v \, dt. \tag{4.1}$$

$v(p, T, c) = v_{i+} - v_{i-}$ is the net flux of the particle i through the surface. If bulk diffusion is rate determining, the shape of a sample must be known to determine the equations for the rate and time laws by appropriate solution of the diffusion problem. The rate and time laws presented in Sects. 4.2, 3 are valid for the limits where only one of the partial steps 1 to 4 mentioned above is rate determining.

4.2.2 Transport of H_2 Molecules in the Gas Phase

The forward reaction rate v_{1+} for this limit is given mainly by the impinging rate of gas molecules on a surface.

$$v_{\max} = \alpha\, 2.7 \cdot 10^{29}\, \frac{p[\text{bar}]}{\sqrt{MT}} \qquad [\text{molecules m}^{-2}\text{s}^{-1}] \tag{4.2}$$

with M molecular weight of the gas molecule. This largest value possible for the reaction rate is reduced if the accommodation coefficient of the gas molecules on the surface, α, is smaller than one. That means physisorption cannot occur in any case for some reason. Usually α is not much less than one for reactive gases on unpoisoned metal surfaces. Another reducing factor is the coverage of the surface by physisorbed gas molecules, $\theta_{H_2(\text{phys})}$. This effect is represented here by the simple term $(1 - \theta_{H_2(\text{phys})})$. Thus, the reaction rate is written as

$$v_{1+} = \alpha v_{\max}(1 - \theta_{H_2(\text{phys})}) = p_{H_2}(1 - \theta_{H_2(\text{phys})})\, k_{1+} \tag{4.3}$$

with $k_{1+} = 2.7 \cdot 10^{10} \cdot \alpha/\sqrt{MT}$ $[\text{ML s}^{-1}]$.
The rate law then becomes

$$\frac{dc}{dt} = 2\frac{S}{V} k_{1+} p_{H_2}(1 - \theta_{H_2(\text{phys})}) \tag{4.4}$$

and the time law

$$c - c_0 = \frac{2S}{V} \int_{t_0}^{t} k_{1+} p_{H_2}(t)\,[1 - \theta_{H_2(\text{phys})}(t)]\, dt \tag{4.5}$$

or in the limit $\theta_{H_2(\text{phys})} \ll 1$

$$\frac{dc}{dt} = 2\frac{S}{V} k_{1+}\, p_{H_2} = \text{const}\, p_{H_2} \tag{4.6}$$

and for constant pressure the time law reads as

$$c = c_0 + \text{const} \cdot p_{H_2} \cdot t. \tag{4.7}$$

The hydrogen absorption kinetics is frequently measured by the pressure change in a closed reactor with constant volume V_R. The concentration change dc is then proportional to the pressure change, $dc = a\, dp$. The factor a depends on the volumes of the receiver and the metal sample. Equation (4.6) is then written as

$$\frac{dp}{p} = -2\frac{S}{V \cdot a} k_{1+}\, dt = \text{const}\, dt \tag{4.8}$$

and the time law becomes

$$\ln\frac{p}{p_0} = -\,\text{const.}\, t \tag{4.9}$$

or

$$p = p_0 \exp\left(-\,\text{const}\, t\right). \tag{4.10}$$

4.2.3 Desorption of Physisorbed Molecules

The rate law for the initial stage of the desorption process is given by

$$\frac{dc}{dt} = -\frac{2S}{V}k_{-1} \cdot \theta_{H_2(phys)} = \text{const} \cdot c^2.$$
(4.11)

Since equilibrium shall exist between $H_2(phys)$ and $H(in\ M)$ $\theta_{H_2(phys)} = \text{const} \cdot c^2$ holds for $\theta_{H_2(phys)} \ll 1$ and $c \ll c_{max}$. The time law then becomes

$$\frac{1}{c} - \frac{1}{c_0} = \text{const} \cdot t.$$
(4.12)

The rate constant k_{1-} is normally a large quantity and desorption of physisorbed molecules is in systems of practical interest not rate determining. If the concentration in the sample approaches the equilibrium value, given by the Sieverts' constant

$$K_S = \frac{c_H}{\sqrt{p_{H_2}}}.$$
(4.13)

The physorption equilibrium is established, too, and the overall reaction rate $v = v_{1+} - v_{1-}$ becomes zero. This yields for the constant in (4.12), $\text{const} = k_{1+}(1 - \theta_{H_2(phys)})/K_S^2$.

4.2.4 Dissociation of Physisorbed H_2 Molecules

In models where the dissociation of adsorbed H_2 molecules is rate determining the physisorption reaction is assumed to be in equilibrium,

$$\theta_{H_2(phys)} = \frac{K_{phys} \cdot p_{H_2}}{1 + K_{phys} \cdot p_{H_2}}$$
(4.14)

or for the low coverages normally expected at not too low temperatures

$$\theta_{H_2(phys)} = K_{phys} \cdot p_{H_2}.$$
(4.15)

The rate law is then written as $dc/dt = v_{2+}$ with

$$v_{2+} = \frac{2Sk_{2+}}{V} \cdot \theta_{H_2(phys)}(1 - \theta_{H(chem)})^2$$
(4.16)

or

$$v_{2+} = \frac{2Sk_{2+}^0 \exp(-A_{2+}/kT)}{V} \cdot \frac{p_{H_2}K_{phys}}{1 + p_{H_2}K_{phys}} \cdot (1 - \theta_{H(chem)})^2.$$
(4.17)

For low pressures $p \ll 1/K_{phys}$ and low coverages $\theta_{H(chem)}$ we get

$$v_{2+} = \frac{2Sk_{2+}^0 \exp(-A_{2+}/kT)}{V} \cdot p_{H_2} \cdot K_{phys}$$
(4.18)

or

$$v_{2+} = \text{const} \cdot p_{H_2} \cdot \exp(-A_{2+}/kT).$$
(4.19)

The time law

$$c - c_0 = \frac{2S}{V} \int_0^t v_{2+} \, dt \tag{4.20}$$

for constant pressure p_{H_2} and $\theta_{H(chem)} \ll 1$ can be written as

$$c - c_0 = \frac{2S}{V} \, \text{const} \cdot p_{H_2} \cdot \exp\left(-A_{2+}/kT\right) \cdot t. \tag{4.21}$$

In this limit a linear time law is obtained, too, and the concentration increase is proportional to the hydrogen pressure. The pressure decrease in a closed reactor with constant volume is described again by the logarithmic time law (4.10),

$$p = p_0 \exp\left(-\text{const} \cdot t\right). \tag{4.22}$$

4.2.5 Recombination and Desorption of Chemisorbed Hydrogen Atoms

The desorption of hydrogen atoms from the chemisorbed state on the surface is controlled by the backward reaction of partial step 2: $dc/dt = -v_{2-}$ with

$$2v_{2-} = \frac{S k_{2-}^0 \exp\left(-A_{2-}/kT\right)}{V} \cdot \left(1 - \theta_{H_2(phys)}\right) \cdot \theta_{H(chem)}^2. \tag{4.23}$$

Equilibrium shall exist between dissolved hydrogen atoms and chemisorbed hydrogen atoms

$$\frac{\theta_{H(chem)}}{\left(1 - \theta_{H(chem)}\right)} = \frac{\theta_{H(m)}}{\left(1 - \theta_{H(m)}\right)} \cdot K_3. \tag{4.24}$$

For $\theta_{H(chem)} \ll 1$, $\theta_{H(m)} \ll 1$, with $\theta_{H(m)} = N_{H(m)}/N_{\text{interstitial sites}}$ defined as areal concentration of hydrogen atoms in interstitial sites of the metal phase (4.24) reads as

$$\theta_{H(chem)} = \theta_{H(m)} K_3 = c K_3^*. \tag{4.25}$$

With $\theta_{H_2(phys)} \ll 1$ and (4.25), Eq. (4.23) can be written as

$$\frac{dc}{dt} = -\frac{S}{2V} c^2 \cdot K_3^* \cdot k_{2-}^0 \exp\left(-A_{2-}/kT\right) \tag{4.26}$$

or

$$\frac{dc}{dt} = -\text{const} \cdot c^2 \cdot \exp\left(-A_{2-}/kt\right). \tag{4.27}$$

The activation energy A_{2-} is not a totally independent quantity. It is determined by the sum of the chemisorption enthalpy of one g-atom hydrogen atoms on the surface, ΔH_{chem}, and the activation energy A_{2+} of the chemisorption forward reaction (Fig. 2.3),

$$A_{2-} = \Delta H_{chem} + A_{2-}. \tag{4.28}$$

The time law is derived from (4.27),

$$\frac{1}{c} - \frac{1}{c_0} = \text{const} \cdot \exp\left(-A_{2-}/kT\right) \cdot t. \tag{4.29}$$

4.2.6 Surface Penetration, Forward Reaction

The transition from the chemisorbed state to the first subsurface layer in the bulk is a process comparable to a diffusion jump in the bulk. Since the activation energies of both processes can differ, introducing surface penetration as a separate partial step is sometimes an advantage. The forward reaction can be written as

$$\frac{dc}{dt} = \frac{S k_{3+}^0 \exp\left(-A_{3+}/kT\right)}{V} \cdot \left(1 - \theta_{H(m)}\right) \cdot \theta_{H(chem)}. \tag{4.30}$$

With the equilibrium condition

$$\theta_{H(chem)} = \frac{\sqrt{p_{H_2}} K_{chem}}{1 + \sqrt{p_{H_2}} K_{chem}} \tag{4.31}$$

for the reaction

$$H_2 + 2\,\text{sites} \leftrightarrows 2\,H(\text{chem}),$$

(4.30) is changed into

$$\frac{dc}{dt} = \frac{S k_{3+}^0 \exp\left(-A_{3+}/kT\right)}{V} \cdot \left(1 - \theta_{H(m)}\right) \cdot \frac{\sqrt{p_{H_2}} K_{chem}}{1 + \sqrt{p_{H_2}} K_{chem}} \tag{4.32}$$

and in the limit $\theta_{H(m)} \ll 1$, $\sqrt{p_{H_2}}\, K_{chem} \ll 1$ into

$$\frac{dc}{dt} = \frac{S k_{3+}^0 \exp\left(-A_{3+}/kT\right)}{V} \cdot \sqrt{p_{H_2}} \cdot K_{chem} \tag{4.33}$$

or

$$\frac{dc}{dt} = \text{const} \cdot \sqrt{p_{H_2}} \cdot \exp\left(-A_{3+}/kT\right). \tag{4.34}$$

The time law for constant pressure is given by

$$c - c_0 = \text{const} \cdot \sqrt{p_{H_2}} \cdot \exp\left(-A_{3+}/kT\right) \cdot t. \tag{4.35}$$

Here, for the first time, the forward reaction is not proportional to the pressure p_{H_2} but to $\sqrt{p_{H_2}}$ because of the chemisorption equilibrium, $\theta_{H(chem)} \propto \sqrt{p_{H_2}}$. The condition for gas absorption in a closed receiver of constant volume yield for the rate law

$$\frac{dp}{\sqrt{p}} = \text{const} \cdot dt \tag{4.36}$$

and for the time law

$$\sqrt{p} = \sqrt{p_0} - 1/2 \cdot \text{const} \cdot t. \tag{4.37}$$

4.2.7 Surface Penetration, Backward Reaction

The backward reaction is given as

$$\frac{dc}{dt} = \frac{-Sk_{3-}^0 \exp\left(-A_{3-}/kT\right)}{V} \cdot \theta_{H(m)} \cdot \left(1 - \theta_{H(chem)}\right). \tag{4.38}$$

In the limit $\theta_{H(chem)} \ll 1$ and $\theta_{H(m)} \propto c$ a rate law

$$\frac{dc}{dt} = -\text{const} \cdot c \cdot \exp\left(-A_{3-}/kT\right) \tag{4.39}$$

and a time law

$$\ln c/c_0 = -\text{const} \cdot \exp\left(-A_{3-}/kT\right) \cdot t \tag{4.40}$$

are obtained. The activation energies A_{3+} and A_{3-} are coupled by the enthalpy ΔH_{seg} of the segregation reaction H(chem) = H(in M) (Fig. 2.3),

$$A_{3+} = A_{3-} + \Delta H_{seg} = A_{3-} \Delta H_{chem} + \Delta H_{sol}. \tag{4.41}$$

4.2.8 Diffusion in the Metal Phase

In the limit where diffusion of hydrogen in the sample becomes rate determining, equilibrium is established between H_2 molecules in the gas phase and hydrogen atoms in the first subsurface layer in the bulk. In an absorption experiment under those conditions the concentration at the surface of the bulk is determined by Sieverts' law

$$c(x = 0, t) = \sqrt{p_{H_2}(t)} \cdot K_S(T) \tag{4.42}$$

and it is assumed to be zero or very low in desorption runs.

In contrast to the limits with surface control, the concentration in the bulk is no longer constant but an inhomogeneous distribution of hydrogen atoms exists with steep gradients at the beginning of an experiment. This distribution and the total amount of absorbed hydrogen atoms after a period of time depends on the shape of the sample. The following questions are normally of practical interest:

i)   How much hydrogen can be absorbed by a sample at given pressure and temperature? This maximum amount is given by the Sieverts' law.

ii)   What is the reaction time needed to attain 90 or 99% saturation of the sample? An exact answer can be given only after a careful analysis of the specific diffusion problem under consideration.

iii)   How thick is the diffusion zone with a high hydrogen concentration near the surface after a given annealing time? Here the approximations for a plane semi-infinite sample can be used for rather reliable estimates (Fig. 2.5).

A detailed discussion of solutions for specific diffusion problems goes beyond the scope of this overview. The reader is referred to the relevant literature, especially, to the excellent reference book of *Crank* [4.25] which offers solutions for many standard problems. The results of diffusion problems in samples with simple geometric shape are frequently given by error functions which contain $\xi = x/2\sqrt{Dt}$ as variable. Therefore, the shape of diffusion profiles can be plotted in terms of this combined time and distance parameter. Normally a diffusion process is almost finished when ξ is about one. Thus, the diffusion time needed to reach an almost homogeneous concentration distribution over a typical distance of x, for example, over the thickness of a sheet or the radius of a sphere, is roughly given by

$$t \approx x^2/(4D) \tag{4.43}$$

with D diffusion coefficient. Typical results for a few standard diffusion problems are presented in Figs 2.5–7.

4.3 Hydride Formation

4.3.1 Reaction Mechanism, Partial Steps

In models where adherent hydride layers are formed on a metal surface more re-action partial steps are involved than in the hydrogen solution models. At least seven processes are active in the reaction,

$$x/2\,H_2(gas) + M \leftrightarrows MH_x;$$

1) $H_2(gas) \leftrightarrows H_2(physisorbed\ on\ MH_x)$,
2) $H_2(phys) \leftrightarrows 2\,H(chemisorbed\ on\ MH_x)$,
3) $H(chem) \leftrightarrows H(in\ MH_x)$,
4) diffusion of H through the MH_x layer,
5) $M + x\,H(in\ MH_x) \leftrightarrows MH_x$,
6) $H(in\ MH_x) \leftrightarrows H(in\ M)$,
7) diffusion of H in M.

The steps 1 to 3 are the same partial processes as for the solution reaction. In both cases the H_2 molecule is physisorbed on the surface, then dissociated to hydrogen atoms and chemisorbed on the outer sample surface. From this state the hydrogen atom jumps to the first subsurface layer inside the hydride scale. In the metal-hydrogen systems it is assumed that the mobile species in the hydride phase are either vacancies on hydrogen lattice sites or hydrogen interstitials. They are responsible for a flux of hydrogen atoms from the surface to the hydride/metal interface. There, with metal atoms from the α phase, new hydride lattice mole-cules are formed (Fig. 4.3).

At the reaction front the hydrogen concentration in the metal phase is normally determined by the terminal solubility or, in other words, the α phase is saturated there with hydrogen atoms. If equilibrium between the α phase and the β phase is attained only at the interface a concentration gradient exists toward the center of the sample. Part of the hydrogen flux through the hydride layer enters then the α phase and diffuses from the interface into the bulk (steps 6 and 7) and the remain-ing part of the hydrogen flux increases the thickness of the hydride layer. Thus, at this interface two processes can proceed in parallel. The hydrogen diffusion in the α metal phase is often for thin samples relatively fast and the equilibrium concen-tration in the α phase is attained before a hydride layer of detectable thickness is formed.

In the models of this section the solution reaction will not be considered. It is assumed that the hydrogen solubility in the metal phase is very small and the solution reaction is very fast when compared with the hydride formation rate. The H_2 pressure must, of course, be higher than the decomposition pressure of the hydride in the absorption runs, and will be very low during desorption. Again, the shape of the sample does not play any role as long as partial steps on the hydride surface are rate determining. Only the ratio surface to volume enters the equations of rate and time laws as a constant factor.

Decomposition of a hydride particle is a totally different process. An α phase metal layer is formed on the surface and diffusion of hydrogen through the hydride phase cannot become rate determining. Therefore, the kinetics of decomposition processes has to be discussed as a separate problem.

4.3.2 Physisorption and Chemisorption

The situation is exactly the same as for the solution reaction and the relations (4.1–29) hold in this limit as well. In the present section, only the mechanism of hydride decomposition shall be discussed for the limit of rate control by recombination of chemisorbed hydrogen atoms on the surface of the metal layer present on the sample surface. The hydrogen concentrations in the metal layer and in the hydride at the hydride/metal interface are fixed by the two phase equilibrium values (Fig.4.7). Consequently, if the diffusion in the α phase and the surface penetration step are assumed to be fast processes the hydrogen concentration in the subsurface position of the metal layer and the chemisorption concentrations $\theta_{H(chem)}$ have fixed values, determined by the hydrogen equilibrium concentration in the α phase of the two-phase sample. Then the rate of the backward chemisorption reaction $v_{2-} = k_{2-}\,\theta_{H(chem)}^2(1 - \theta_{H_2(phys)})$ must have a constant value, too. In equilibrium the rates for the forward and for the backward reaction are equal and the plateau pressure, $p_{H_2(plateau)}$, is established. Therefore, if the bulk of a sample is still in the two phase region and in chemical equilibrium; and if in addition, the physisorption equilibrium is established, the desorption rate can be written as

$$v = v_{2-} - v_{2+} = v_{2-}(1 - v_{2+}/v_{2-}), \tag{4.44}$$

where v_{2+} and v_{2-} are given by (4.16, 23) respectively. The expression $\theta_{H_2(phys)}/(1 - \theta_{H_2(phys)})$ in the term v_{2+}/v_{2-} can be replaced by $p_{H_2}K_{phys}$ and all other terms are constant under the assumptions made and we get $v_{2+}/v_{2-} = \text{const}\,p$. The equilibrium condition $v_{2+}/v_{2-} = 1$ is obeyed for $p_{H_2} = p_{H_2(plateau)}$. Since v_{2-} is constant the degassing rate can be written as

$$v = \frac{S}{V}\,\text{const}\left(1 - p_{H_2}/p_{H_2(plateau)}\right). \tag{4.45}$$

4.3.3 Transition from Chemisorption to the Hydride Subsurface

Two types of defect models can be used to describe the activity of hydrogen atoms in the hydride phase of an ideal system.
i) Hydrogen atoms shall be incorporated only in regular lattice sites of the hydride phase. Then vacant hydrogen lattice sites are needed as migrating particles

of the transport mechanism with an areal concentration $\theta_{H(vac)} = (1 - c/c_{stoichiometric})$. This is a realistic model for hydrides with understoichometric hydrogen concentration at the end of the pressure plateau.

ii) Hydrogen is incorporated in the hydride in interstitial sites. Then the concentration of migrating interstitials needed for the transport process is given by a term $c_{H(int)} = c - c_{stoichiometric}$. The concentration of the hydride at the end of the plateau must then be higher than the value of the stoichiometric compound. The areal concentration $\theta_{H(int)} = c_{H(int)}/zc_{stoichiometric}$ is defined by the ratio z of interstitial sites to regular lattice sites for hydrogen in the hydride.

Applying the same treatment as for hydrogen solutions [Sect. 4.2.6, (4.32, 35)] yields

$$\frac{dc}{dt} = \frac{Sk_{3+}^0 \exp\left(-A_{3+}/kT\right)}{V} \cdot \theta_S \cdot \frac{\sqrt{p_{H_2}}K_{chem}}{1 + \sqrt{p_{H_2}}K_{chem}} \tag{4.46}$$

with θ_S areal concentration of empty sites for atoms in the hydride, $\theta_{H(vac)}$ or $1 - \theta_{H(int)}$). For the interstitial model with $\theta_S = 1$ and $\theta_{H(chem)} \ll 1$ we get

$$c - c_0 = \text{const} \cdot \sqrt{p_{H_2}} \cdot \exp\left(-A_{3+}/kT\right) \cdot t. \tag{4.47}$$

If this partial step is rate determining it can be assumed again that the backward reaction rate for two phase samples is desorption from the saturated α phase with a constant rate. This rate must be equal to the forward reaction rate at the plateau pressure. By a similar treatment as mentioned above the expression for the rate law is given as

$$v = \frac{S}{V}\text{const}\left(1 - \sqrt{p_{H_2}/p_{H_2(\text{plateau})}}\right). \tag{4.48}$$

4.3.4 Diffusion in the Hydride Phase

The diffusion flux of a species is given by Ficks' first law

$$J = -bc\nabla\mu, \tag{4.49}$$

where c and μ are concentration and chemical potential of the moving defect, respectively, and b the mobility. The migrating defects in the hydride phase are normally vacant hydrogen lattice sites if the hydrogen concentration is smaller than the stoichiometric concentration or hydrogen interstitial atoms with a concentration $c_{H(int)}$ if the hydrogen concentration is higher. The chemical potential μ_{defect} can be written as

$$\mu = \mu^0 + kT \ln a, \tag{4.50}$$

where a is the activity of the defect. For the vacancy mechanism the defect formation reaction can be written as (Appendix A)

$$1/2\,H_2 + \text{vacancy} = 0,$$

$$a_v = K_v(T)/\sqrt{p_{H_2}} = \gamma \cdot c_v \tag{4.51}$$

If vacancies can be described by the equations for diluted systems, γ is a constant. This yields for the flux J an expression

$$J = -b\frac{a_v}{\gamma} \cdot kT\frac{1}{a_v}\nabla a_v \tag{4.52}$$

or

$$J = -\frac{b}{\gamma}kT\nabla\frac{K_v}{\sqrt{p_{H_2}}}. \tag{4.53}$$

As long as vacancies in the hydride are treated as a diluted component, the constant activity coefficient γ can be set to unity, and (4.52) can be written as

$$J_H = -J_v = D \cdot \nabla a_v = D_v \cdot K_v \cdot \nabla(1/\sqrt{p_{H_2}}). \tag{4.54}$$

For hydrogen interstitials as migrating defects the reaction equation

$$H_2 + \text{site} = H(\text{interstitial})$$

holds and

$$a_i = \sqrt{p_{H_2}} \cdot K_i(T) = \gamma\, c_i, \tag{4.55}$$

$$J_H = J_i = -D\nabla a_i = -DK_i\nabla\sqrt{p_{H_2}}. \tag{4.56}$$

The mathematical treatment of diffusion-controlled processes has to consider the shape of the sample, and different results are obtained for individual models. Here only two cases shall be discussed, namely, the plane geometry of sheets and the spheric geometry as prototype of powder particles. The relations discussed below refer to the formation of a hydride layer during gas absorption. If a desorption process is controlled by diffusion of hydrogen interstitial atoms through a growing metal surface layer, the relevant formulas can be used as well. If the sign of the activity gradient is changed the direction of the diffusion flux is inverted, too. Only in the equations correlating the flux and the layer growth a negative sign has to be inserted.

Plane samples
According to Fick's first law the growth of a hydride layer can be described by the equation

$$\frac{dl}{dt} = RJ = \pm R \cdot D\nabla a = -RD\frac{\Delta a}{l} \tag{4.57}$$

with the layer thickness l [m], the hydrogen flux J [ML/s], the conversion factor R [m/ML] and D [m^2/s]. The quantity a is defined as the activity of the migrating hydrogen defect in the hydride. As mentioned above, $c = a$ holds since γ can be set one by definition for an ideal diluted system. As long as the sample consists of two phases, the surface of the hydride phase is in this approximation in equilibrium with the gas phase and the internal interphase in equilibrium with the α phase. Then, the defect concentration at the surface becomes

$$c_{H(vac)} = K_v/\sqrt{p_{H_2}} \tag{4.58}$$

for vacancies and

$$c_{H(int)} = \sqrt{p_{H_2}} \cdot K_i \tag{4.59}$$

for interstitials. At the hydride/metal interface the pressure terms $\sqrt{p_{H_2}}$ in (4.58) and (4.59) have to be replaced by $\sqrt{p_{H_2(plateau)}}$ that means by the value of the two phase equilibrium pressure. With these data the rate law for vacancies can be written as layer growth law

$$\frac{dl}{dt} = -\frac{RDK_v}{l}\left(\frac{1}{\sqrt{p_{H_2}}} - \frac{1}{\sqrt{p_{H_2(plateau)}}}\right) \tag{4.60}$$

and for the time law we get

$$l^2 = -\frac{R}{2}DK_v\left(\frac{1}{\sqrt{p_{H_2}}} - \frac{1}{\sqrt{p_{H_2(plateau)}}}\right)t. \tag{4.61}$$

For interstitials the equations read as

$$\frac{dl}{dt} = \frac{RDK_i}{l}\left(\sqrt{p_{H_2}} - \sqrt{p_{H_2(plateau)}}\right) \tag{4.62}$$

and

$$l^2 = \frac{R}{2}DK_i\left(\sqrt{p_{H_2}} - \sqrt{p_{H_2(plateau)}}\right)t. \tag{4.63}$$

Spheric samples
Another type of the samples frequently used in models and accessible to mathematical treatment in closed form are particles with spheric geometry [4.26]. The quasi-stationary flux approximation implies that the flux over all virtual spheric surfaces inside the sample must be constant.

$$\frac{J}{r^2} = -4\pi D \frac{da}{dr}. \tag{4.64}$$

This yields for vacancies

$$J\int_{r_\alpha}^{r_0}\frac{dr}{r^2} = -4\pi D \int_{\alpha_{r_a}}^{\alpha_{r_0}} da = 4\pi D\left(\frac{1}{\sqrt{p_{H_2}}} - \frac{1}{\sqrt{p_{H_2(plateau)}}}\right); \tag{4.65}$$

after integration on the left-hand side, the flux of vacancies can be written as

$$J_v = 4\pi D\frac{\dfrac{1}{\sqrt{p_{H_2}}} - \dfrac{1}{\sqrt{p_{H_2(plateau)}}}}{\dfrac{1}{r_0} - \dfrac{1}{r_\alpha}}. \tag{4.66}$$

If the shape of samples for a specific system and the experimental procedure are known, the flux J can be expressed as a function of the radius of the reaction front,

r_α. For an experiment performed under constant pressure p_{H_2} a growth law for the layer thickness is obtained of the form

$$4\pi r_\alpha^2 \frac{dr_\alpha}{dt} = RJ(p_{H_2}, p_{H_2(\text{plateau})}, r_0, r_a) \tag{4.67}$$

and by integration together with (4.66) a time law

$$\int_{r_0}^{r_\alpha} \left(r_a - \frac{r_\alpha^2}{r_0} \right) dr_\alpha = R \cdot J(p_{H_2}) \cdot t \tag{4.68}$$

or

$$(r_0^2 - 3r_\alpha^2 + 2r_\alpha^3/r_0) = -24\pi RD \left(\frac{1}{\sqrt{p_{H_2}}} - \frac{1}{\sqrt{p_{H_2(\text{plateau})}}} \right) t. \tag{4.69}$$

In models with an interstitial migration mechanism the term $1/\sqrt{p_{H_2}} - 1/\sqrt{p_{H_2(\text{plateau})}}$ in (4.69) is replaced by the term $\sqrt{p_{H_2}} - \sqrt{p_{H_2(\text{plateau})}}$.

If the gas absorption of N spheres in a closed receiver of the volume V is described by a model, the amount of absorbed hydrogen, n_H in moles, the pressure change dp, and the layer thickness dr are interrelated by the equation

$$V\,dp_{H_2} = N \cdot k \cdot 4\pi r^2\,dr = K\,dn_H \tag{4.70}$$

with K and k being constants. With these procedures quantitative model calculations on diffusion-controlled hydrogen absorption in isobaric or isochoric systems can be performed. The time laws are obtained by integration of the rate laws given, where necessary by computer aid [4.24].

4.3.5 Formation of the Hydride Phase

In this limit an interface reaction is rate controlling. The reaction takes place at the hydride/metal interface and can be written for the vacancy model as

$$M \leftrightarrows MH_x + xH(\text{vac}),$$

$$\frac{a_{MH_x} \cdot a_{H(\text{vac})}^x}{a_M} = K(T) \text{ or } a_{H(\text{vac})} = K^*(T). \tag{4.71}$$

The forward reaction rate is constant

$$v_{4+} = k_{4+}(T) \tag{4.72}$$

and for the backward reaction one can write

$$v_{4-} = a_{H(\text{vac})}^x \cdot k_{4-}(T). \tag{4.73}$$

The rate of decomposition of the hydride phase as backward reaction must again be equal to the forward reaction rate if $a_{H(\text{vac})}$ has reached the equilibrium concentration defined by the maximum concentration at the end of the pressure plateau

(Fig. 4.7). Thus, the rate of the forward net reaction rate can be written as $v = v_{4+} - v_{4-}$ or

$$v_4 = k_{4+}\left[1 - (a_{H(vac)}/a_{H(vac,equilibrium)})^x\right].$$

(4.74)

For the vacancy-type mechanism $a_{H(vac)} = K(T)/\sqrt{p_{H_2}}$ holds and one can write

$$v = k\left[\left(\frac{K(T)}{\sqrt{p_{H_2(plateau)}}}\right)^x - \left(\frac{K(T)}{\sqrt{p_{H_2}}}\right)^x\right]$$

(4.75)

and

$$\frac{dc}{dt} = F(r_\alpha) \cdot v = F(r_\alpha) \cdot k\left[\left(\frac{K(T)}{\sqrt{p_{H_2(plateau)}}}\right)^x - \left(\frac{K(T)}{\sqrt{p_{H_2}}}\right)^x\right]$$

(4.76)

with $F(r_\alpha)$ area of the metal/hydride interface. For the interstitial-type mechanism the reaction equation reads as

$$x\,H(int) + M = MH_x$$

and by an analogous procedure we get with (4.55)

$$\frac{dc}{dt} = F(r_\alpha) \cdot v = F(r_\alpha) \cdot k\left\{\left[K(T)\sqrt{p_{H_2}}\right]^x - \left[K(T)\sqrt{p_{H_2(plateau)}}\right]^x\right\}.$$

(4.77)

Here the forward reaction rate depends on the activity of the hydrogen interstitials. For samples with plane geometry $F(r_\alpha)$ is a constant of the system. However, if the system consists of N spheres with a constant diameter $2r_0$ the area of the reaction front decreases with increasing layer thickness proportional to the amount n of absorbed H_2 molecules. That yields

$$F(r_\alpha) = 4\pi r_\alpha^2 = 4\pi r_0^2\left(1 - \frac{3k \cdot n_H}{N 4\pi r_0^3}\right)^{2/3}.$$

(4.78)

The constant k correlates the moles of hydrogen absorbed by hydride formation with the geometric factors of the model.

4.4 Computer Simulation of Advanced Models

The models based on the conception of a single rate determining step contain some rather restricting assumptions. Whether or not the approximations applied are justified in a system of interest, and yield realistic rate and time laws, can be studied by more advanced models which incorporate less severe confinements in the mechanisms. Such a model has to combine the forward and backward reaction rates of several, if possible of all, individual partial steps. Furthermore, at any site of the model the equation of continuity

$$\frac{\partial c}{\partial t} = \nabla J + \sum \text{concentration changes by reactions}$$

(4.79)

must be obeyed. This means that the change of the local hydrogen concentration $\partial c/\partial t$ is caused either by a divergence of the flux vector or by a chemical reaction. If the individual partial steps and the diffusion fluxes in the α and β phase are linked by this condition, the hydride layer growth and the concentration profiles in both phases can be determined as a function of pressure and exposure time. The structure of such models define a large number of parameters and constants, and numerical methods must be applied to obtain the results desired. In addition to the experimental parameters H_2 pressure, temperature, and the dimensions of the sample, also the rate constants and activation energies of the individual partial reactions and the diffusion constants for hydrogen in the α and β phase must be known. Further variables are the concentrations of physisorbed and chemisorbed hydrogen species on the hydride surface.

A model of this type for a spherical particle has been developed for the simulation of the absorption and desorption kinetics of hydrogen storage materials [4.23, 27]. The reaction mechanisms proposed are the same as shown in Fig. 4.3. The aim of this kind of modeling is less fitting of individual experimental data but more the description and understanding of reaction mechanisms which lie in an intermediate range between the limits characterized by models using one rate-determining step only. The advanced models can show which of the numerous parameters affect the rate and time laws and how much they have to be changed to accomplish a complete transition from one limit to another one. In this section the structure of such a model and some informative results on hydride formation and decomposition will be discussed. They demonstrate the capabilities of the numerical techniques applied and illustrate the role of individual parameters within the reaction mechanism considered. The model itself is restricted to the minimum number of parameters required for a straightforward treatment of the problem. For less-critical partial reactions, equilibrium conditions have been inserted to avoid confusion by a voluminous mathematical structure. More detailed information on definitions and properties of individual partial steps is found in Sects. 2.3, 4.2, 3 and in the Appendix A.2.

4.4.1 Structure of the Model

The reaction mechanism schematically displayed in Fig. 4.3 consists of the partial reactions physisorption, chemisorption, surface penetration, diffusion the β hydride, hydride formation and diffusion in the α-metal phase of the core of a sphere. In Fig. 4.13 the energies of hydrogen atoms in different states along the reaction path are shown. All energy terms are given as distance from the energy level of a hydrogen atom in the H_2 molecule in the gas phase which has been defined to be zero. The meaning of the quantities and reasonable values used for modeling a standard curve are presented in Table 4.1. The model does not consider the following aspects which can strongly affect the hydriding kinetics in real systems:

– *Lattice expansion due to hydride formation:*
  In most systems hydride formation is accompanied by a volume expansion of about 10%. This causes spalling of the hydride scale, crack formation, and reduction of the size of powder particles.
– *Poisoning of the sample surface:*
  This aspect of surface reactions is discussed in Chap. 6. In most real systems surface poisoning plays a crucial role in kinetics. Only few of the data published so far on the kinetics of low-temperature hydrogen absorption can claim

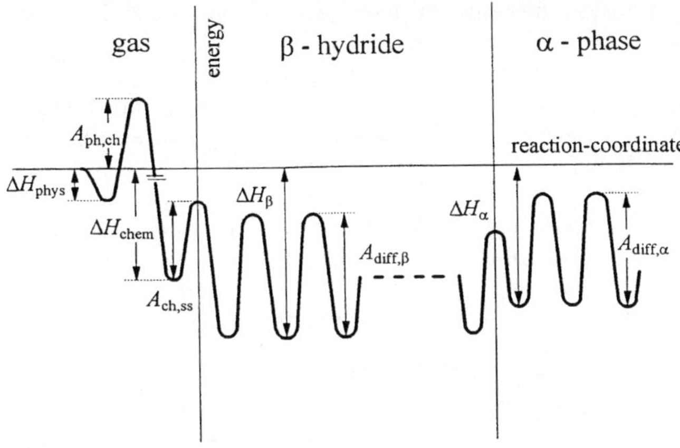

Fig. 4.13 Energetic data of the model on hydride formation, presented in a energy-reaction path diagram

Table 4.1 Standard parameters of the model on hydride formation

parameter	symbol	value	
experimental parameters			
pressure	p	1–50	bar
temperature	T	295	K
powder radius	r_0	1–10	μm
material parameters			
Diffusion constant	$A_{diff,\beta}$	$0.242-0.275$	eV
	$D_{\beta,300K}$	$8\cdot10^{-13}-6\cdot10^{-12}$	m^2s^{-1}
Reaction enthalpies	ΔH_β	$-(0.156-0.175)$	eV
	$\Delta_{H(chem)}$	$-(0.20-0.50)$	eV
	$\Delta_{H_2(phys)}$	$-(0.17-0.30)$	eV
Reaction entropy	$\Delta S_\beta,\ \Delta S_a$	$-5.9\cdot10^{-4}$	eV K^{-1}
Plateau pressure (295K)	$p_{plateau}$	≈ 2.0	bar
H concentration LaNi$_5$H$_6$	$c_{0,\beta}$	$5.6\cdot10^{28}$	H atoms m^{-3}
Aktivation energies			
Chemisorption	$A_{ph,ch}$	< 0.5	eV
Surface penetration	$A_{ch,ss}$	< 0.6	eV
Constants			
Frequency faktor	ν	10^{13}	s^{-1}
Surface monolayer	$N_{ph},\ N_{ch}$	$2.5\cdot10^{19}$	species/m^2

to be absolutely free of errors caused by an oxide skin or other poisons present on the sample surface.

– *Gas phase transport:*
Gas phase diffusion processes can affect the reaction rates, especially in gas mixtures, if powder samples are hydrided at higher pressures, for example, in reaction beds of storage tanks.

– *Thermal effects:*
If the heat of hydride formation is not efficiently removed from the sample, the equilibrium plateau pressure rises due to the increased temperature (Figs. 4.2, 7). The reaction rate is reduced and if the new plateau pressure exceeds the exposure pressure hydride is not formed but decomposed.

– *Hysteresis:*
Thermal effects and internal stresses caused by the volume expansion lead to a hysteresis effect. This causes a difference of the plateau pressure when measured in absorption and desorption runs. It is increased with increasing reaction rate.

– *Grain boundary and structure effects:*
Grain boundary diffusion and the dependence of diffusion coefficients or adsorption processes on the lattice structure and orientation are not considered. The reaction front is assumed to proceed in ideal spherical shells inside the sample.

According to the assumptions made by the model the simulation describes hydride formation of an ideal, unpoisoned spherical sample without volume changes in an isothermal experiment performed in a closed receiver which is loaded to an initial pressure. Isobaric conditions are obtained when the volume is made very large. This problem is known in reaction kinetics as the unreacted core model [4.26]. The forward and backward reaction rates of the physisorption, chemisorption, and the surface penetration reaction are described by the same equations, as discussed in preceding sections.

The diffusion problem is solved by numerical methods using Ficks' second law

$$\frac{\partial c}{\partial t} = D\left(\frac{\partial^2 c}{\partial r^2} + \frac{2\partial c}{r\partial r}\right). \tag{4.80}$$

The sample with a radius r_0 is subdivided into N shells with a thickness $dr_0 = 1/N$. ∂r is replaced by $r_0 - dr_0$ and the position of the shell i inside the layer is then given by $r_i = ir_0 dr_0$. The concentration of the shell i at the time $t + \delta t$ is obtained from the concentrations in the shells $i - 1$, i, $i + 1$ at the time t and the concentration c_i at the time t by the equation [Ref.4.25, p. 149]

$$c_{i,t+\delta t} = c_{i,t} + \frac{\delta t\, D}{i(r_0\, dr_0)^2}[(i + 1)c_{i+1,t} - 2ic_{i,t} + (i - 1)c_{i-1,t}]. \tag{4.81}$$

The p–c isotherms of the type shown in Fig. 4.14 can be simulated best by an activity term for hydrogen in the hydride as defined by the Lacher model [4.28] and [Ref.4.1, pp. 90–117]. The activity is given by the number of empty hydrogen lattice sites in the hydride as a function of concentration, see (4.51).

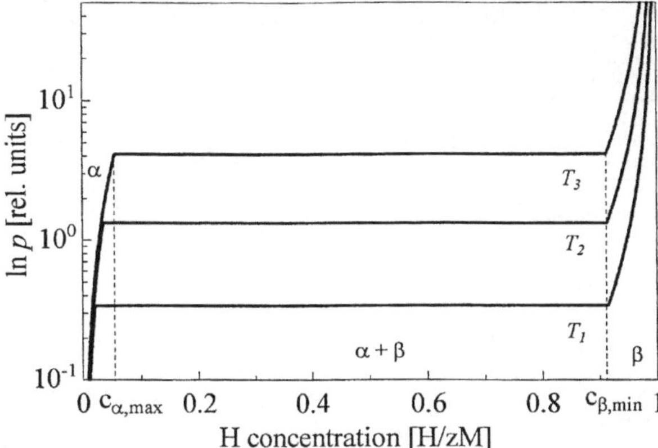

Fig. 4.14 Ideal equilibrium pressure - concentration isotherms of the model system calculated for three different temperatures

The rate of the phase transformation process is approximated by a rate equation

$$v = \text{const}(c - c_{\beta,\text{plateau}}). \tag{4.82}$$

This simple rate equation guarantees that v is zero for $c = c_{\beta,\text{plateau}}$ and becomes faster if the distance of the hydrogen concentration in the β phase deviates from the plateau value. At the hydride surface the difference between the forward and backward surface penetration rates must be balanced by the diffusion flux plus the concentration change at the shell $i = N$. The difference of the diffusion fluxes $D_\beta - D_\alpha$ at the α/β interface determines the hydride layer growth rate. The concentration changes of physisorbed and chemisorbed hydrogen have been neglected in material balances since the gas amount present on the surface is small compared with the total amount of hydrogen absorbed by the bulk.

4.4.2 Procedure of the Numeric Solution

Figure 4.15 exhibits a schematic flux diagram of the procedure for the numeric solution of the mathematical problem. At first the material parameters must be chosen which define a system. Then the experimental parameters for a special run must be defined. These are temperature, pressure, and the initial hydrogen concentration of the sample. After the fixing of constants three subroutines are passed for the determination of concentrations in the surface area. The concentration of physisorbed hydrogen molecules is taken from the equilibrium constant and the gas pressure. The surface coverage with chemisorbed hydrogen atoms is obtained from the steady state condition for the fluxes to the physisorbed and to the subsurface layer using the approximation $\partial c/\partial t = 0$. Then, a value for the time interval δt has to be chosen which is compatible with the requirements of the algorithm and the concentration c_N at the subsurface layer is calculated from the equation of continuity (4.79). At this point the further procedure depends on the value of c_N. If c_N is below the terminal solubility of the α phase no hydride has been formed and a subroutine calculates diffusion in the α phase. If β hydride is

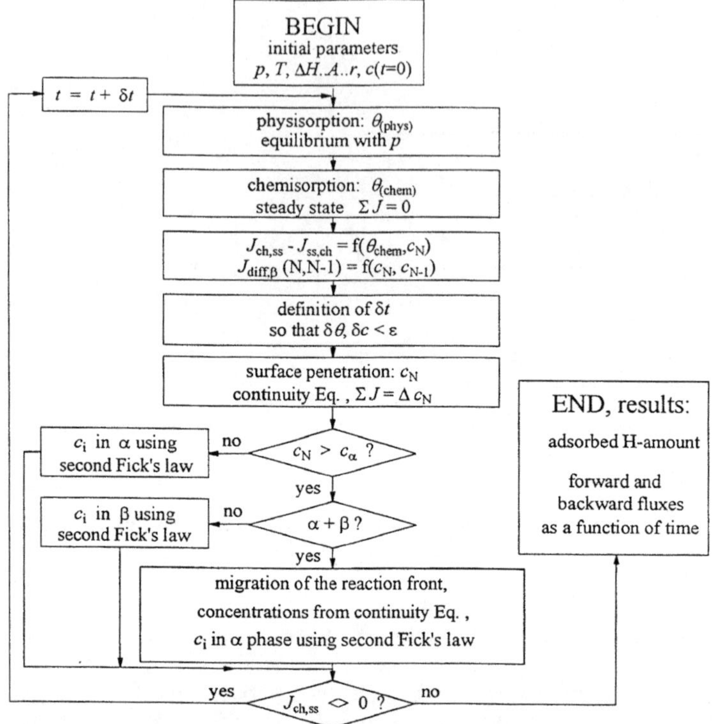

Fig. 4.15 Flux diagram of the numerical procedure for model calculations on hydride formation of spheric particles

already present at the sample surface the program checks whether the sample has still a core consisting of α phase or it has already been totally transformed to hydride versus the end of a run. For two phase samples the thickness of the hydride layer and the concentration profiles in the α and β phase can be calculated for any desired exposure time using the relation (4.81).

Figure 4.16 depicts a set of concentration profiles of simulation runs after different exposure times. In addition to the bulk concentrations, the concentration variables, $\theta_{H_2(phys)}$, $\theta_{H(chem)}$, the hydride layer thickness and the forward and backward flux of each partial step is compiled as a function of time and of the amount M of hydrogen absorbed. This quantity $M = \bar{c}/\bar{c}_{max}$ with $0 < M < 1$ is normalized to the maximum concentration of the sample if it is totally converted to hydride with the equilibrium composition $c_\beta(T, p)$ given by the experimental parameters. The numeric results can be plotted also as curves of the absorbed amount of hydrogen, M, versus exposure time and compared immediately with experimental absorption curves. Curves of this type are shown in Fig. 4.17. The parameters given in Table 4.1 have been kept constant. Only the activation energy of the chemisorption step forward, $A_{ph,ch}$, has been varied. The time required to attain equilibrium increases by two orders of magnitude if $A_{ph,ch}$ is changed by a factor of three. Furthermore, the time law is $M \propto t^{1/2}$ for low and $M \propto t$ for large $A_{ph,ch}$ values.

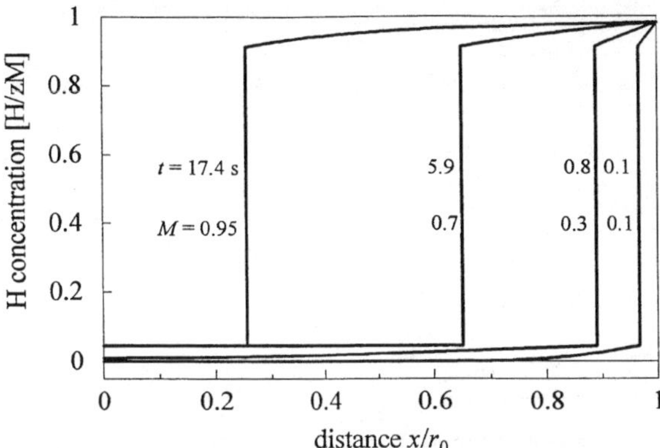

Fig. 4.16 Concentration profiles in a spheric sample during hydriding. The hydride formed is H_2M, and the symbol M represents the the relative degree of hydriding $\bar{c}/\bar{c}_{max}$

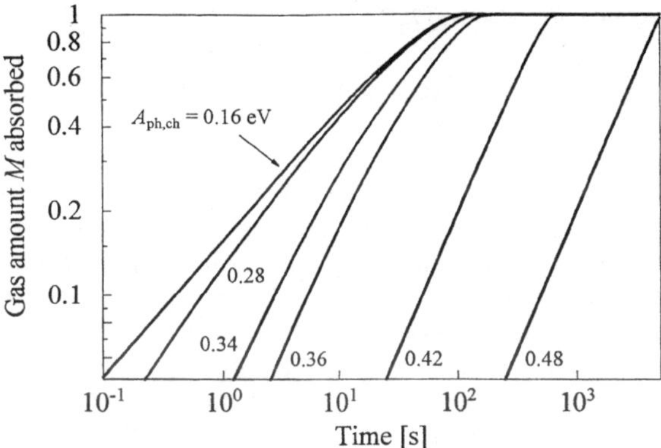

Fig. 4.17 Simulated hydrogen absorption curves for various values of the chemisorption activation energy $A_{ph,ch}$

Because of the large number of parameters available, it is easy to fit any experimental curve. Therefore, it would not make much sense to present here a collection of experimental data together with simulated absorption curves. The benefit of an advanced model is the chance to study the role of individual parameters within the mechanism proposed for the description of an overall reaction. The most valuable quantities for the understanding are the maximum diffusion flux through the β phase, $J_{diff,max}$, if the equilibrium concentrations are established at both boundaries, the forward fluxes for dissociative chemisorption, $J_{ph,ch}$, and for surface penetration, $J_{ch,ss}$. These quantities can be used for the characterization of the behavior of the overall reaction. One can define a variable $f_c = J_{ph,ch}/J_{diff,max}$ which is large for diffusion control and small for chemisorption control. A similar relation $f_s = J_{ch,ss}/J_{diff,max}$ indicates that surface penetration is the rate determining partial step if $f_s \ll 1$ holds.

Valuable information can be obtained if the diffusion flux density in the uppermost shell i = N of the hydride, J_{diff}, the forward flux for chemisorption, $J_{ph,ch}$,

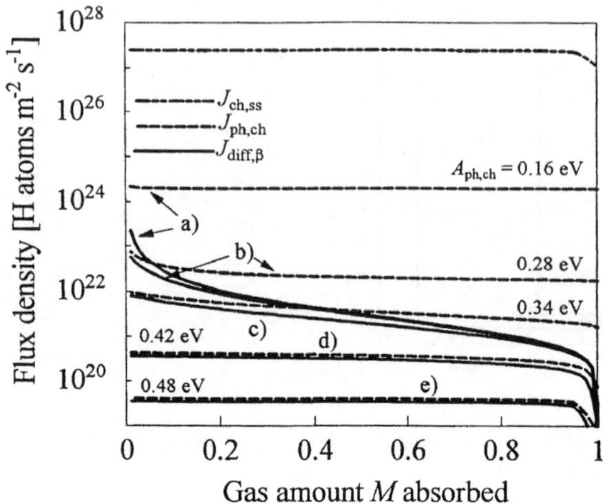

Fig. 4.18 Hydrogen flux densities of the partial fluxes for chemisorption, $J_{ph,ch}$, surface penetration, $J_{ch,ss}$, and diffusion, $J_{diff,\beta}$ for the simulation runs shown in Fig 4.17

and the forward flux for surface penetration, $J_{ch,ss}$, are plotted as a function of the amount of absorbed hydrogen, M. The curves depicted in Fig. 4.18 are taken from the same runs as shown in Fig. 4.17, again for different $A_{ph,ch}$ values as parameter. The surface penetration flux, $J_{ch,ss}$, has a high value. It cannot contribute much to the mechanism since it is in this example a very fast reaction partial step. The same is true for the chemisorption flux, $J_{ph,ch}$ at small $A_{ph,ch}$ values of 0.16 and 0.28 eV. The diffusion flux, which is identical with the reaction rate, is much smaller than the forward chemisorption flux. However, if the activation energy of the chemisorption reaction is increased, the slow rate of this step also reduces the diffusion flux. Diffusion cannot be faster than the supply of atoms passing the chemisorption step or, in other words, the overall reaction rate has to be adapted to this slowest step. In the model this is accomplished by lowering the hydrogen concentration in the subsurface layer, c_N, to a value close to the equilibrium value at the α/β interface. The curves in Fig. 4.18 demonstrate that the run a) is clearly diffusion controlled. The diffusion flux $J_{diff,\beta}$ is here identical with the maximum diffusion flux, $J_{diff,max}$. The runs b) and c) belong to a mixed regime, chemisorption controlled at low mean bulk concentrations at the beginning of the absorption run and more strongly diffusion controlled versus the end of the simulated experiment. The runs d) and e) are chemisorption controlled, almost up to the very end of the experiments.

4.4.3 Discussion of Results for Absorption

It has been demonstrated in previous sections of this chapter that the exponents $0.5 < n < 1$ in the time law $M \propto t^n$ and in the rate law $dM/dt \propto p^n$ indicate which of the partial steps can be rate determining, the processes at the sample surface, at the α/β interface, or transport phenomena inside the α or β phase. This is an interesting question for handling problems in practical application since it is very important to know where the bottleneck of a reaction is located. If it is a surface step one has to think about catalyzers or poisoning reactions must be eliminated. If it is bulk diffusion, then the size of a sample, the microstructure, grain boundary impurities or the defect structure of the hydride phase play a crucial role in kinetics.

The example presented in the Sect. 4.4.2 has demonstrated how a reaction mechanism is changing from diffusion control to chemisorption control if the rate constant $k_{2+} = k_{2+}^0 \cdot \exp\left(-A_{ph,ch}/kT\right)$ is decreased. This transition can be analysed in more detail if the exponent n in the time law is plotted versus the logarithm of the parameters f_c or f_s. The quantity n represented by the ordinate can be checked by experiment. The parameters f_c or f_s, respectively, are the ratios of partial fluxes. They depend on various system parameters and can be replaced by ratios of other quantities of interest as it will be shown below.

Transition from chemisorption control to diffusion control
Figure 4.19 plots the exponent n versus $\log f_c$ with $f_c = J_{ph,ch}/J_{diff,max}$ for $M = 0.25$ (Fig. 4.17). Curve a is a result of numeric model calculations whereas curve b uses the chemisorption forward flux taken from the equation for the limit of chemisorption control (4.18). The difference between the approximate and the exact value for $J_{ph,ch}$ is caused mainly by the the term for the hydrogen concentration in the β phase. The numeric calculation uses correct values whereas in the approximation $c_{\beta,min}$ (Fig. 4.14) has been inserted. The most important message of this presentation of simulated data is the finding that the transition from chemisorption control ($n = 1$) to diffusion control ($n = 0.5$) requires in this example a change of three to four orders of magnitude for the flux ratio represented by the parameter f_c. The width and position of the transition region of the exponent n in terms of real system parameters depend also on the concentration term M. The curves for run c) in Fig. 4.18 demonstrate that the absorption reaction can be surface controlled in the initial part of an experiment and becomes diffusion controlled versus the end. In Fig. 4.20 the curve b of Fig. 4.19 is plotted versus the ratios of some real system parameters. They are the hydrogen diffusion coefficient in the β-hydride phase, D_β, the particle radius, r_0, the activation energy $A_{ph,ch}$ and the occupancy of surface sites by poisoning particles (Chap. 6). The quantities are normalized by terms denoted by zeros. They are chosen so that n becomes 0.75 for

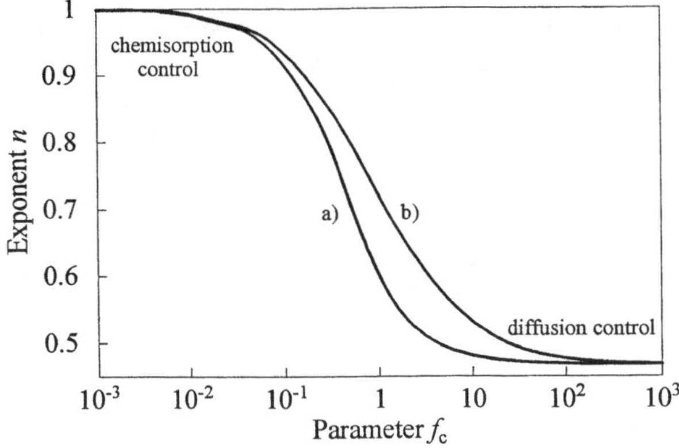

Fig. 4.19 Transition from diffusion control to chemisorption control indicated by the exponent n in the experimental time law $M \propto t^n$ as a function of the rates of partial fluxes $f_c = J_{ph,ch}/J_{diff,max}$

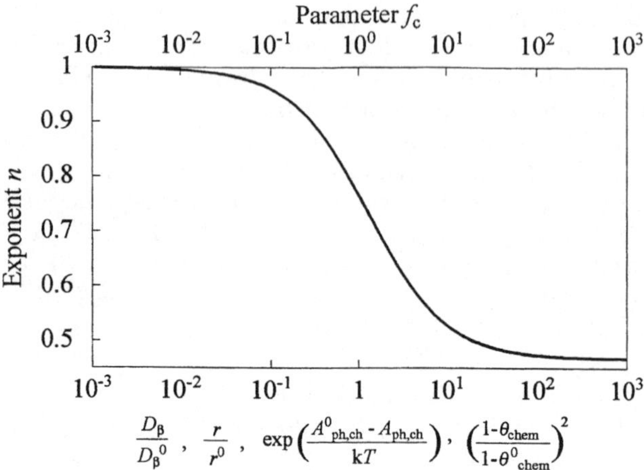

Fig. 4.20 Exponent n of the time law $M \propto t^n$ as a function of system parameters normalized to a value where $n = 0.75$ holds

the zero values. These data demonstrate that the transition from one limiting region to another is smooth and extended transition regions with experimental n values between the limits 1 and 0.5 must be considered as a quite normal result of experimental investigations. A change of the diffusion coefficient or of the mean particle diameter by two to four orders of magnitude can not be normally accomplished by a single experimental technique. Therefore, even a large variation of the parameters available for an experimental investigation sometimes may not be able to cover the whole transition region of a complex mechanism with several rate determining partial steps.

Transition from surface penetration to diffusion control
The pressure dependence of the hydriding rate is here proportional to $p^{1/2}$ for both limits, rate determination by surface penetration or by diffusion. However, the exponent n in the time law remains different. It is one for interface penetration and 1/2 for diffusion control. The curves in Figs. 4.21, 22 are similar to those presented above. They show the competition between surface penetration and bulk diffusion in the hydride as rate determining processes by the change of the exponent n in the time law. The chemisorption step has been made fast compared with the two other partial steps considered. The shape of the curves as well as their interpretation are the same as discussed above.

Experimental data which show in an Arhenius plot reaction rate v versus $1/T$ bended curves are usually interpreted correctly as a transition in the reaction mechanism. But the conclusion, sometimes made, that each end of the curves can be attributed to a simple mechanism for the limit with only one rate determining step is often not justified. The model calculations reveal that the temperature range for a transition from one limit to the other can be quite large if the difference between the two activation energies of interest is small. This is demonstrated in Fig. 4.23. A difference of 0.3 eV or 30 kJ/mol and a temperature change of 150 K is required to accomplish an almost complete change of the reaction mechanism at ambient temperature for the example chosen.

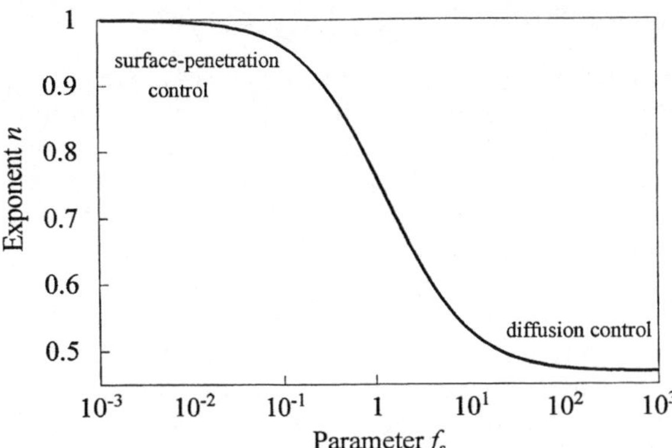

Fig. 4.21 Transition from surface-penetration control to diffusion control indicated by the exponent n in the time law $M \propto t^n$ as a function of the ratio of partial fluxes $f_s = J_{ch,ss}/J_{diff}$

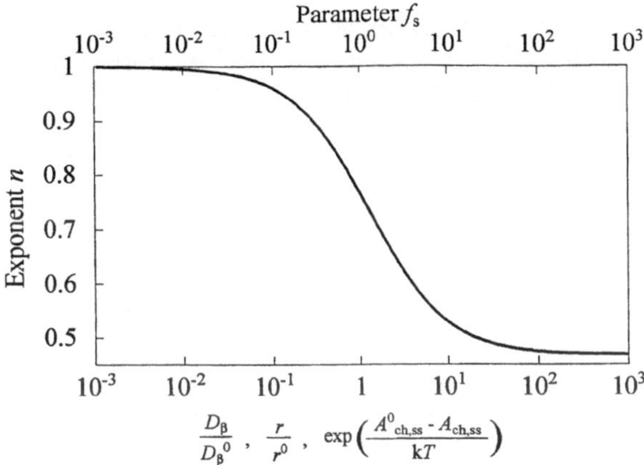

$$\frac{D_\beta}{D_\beta^0} \ , \quad \frac{r}{r^0} \ , \quad \exp\left(\frac{A^0_{ch,ss} - A_{ch,ss}}{kT}\right)$$

Fig. 4.22 Exponent n of the time law $M \propto t^n$ as a function of system parameters normalized to a value where n = 0.75 holds

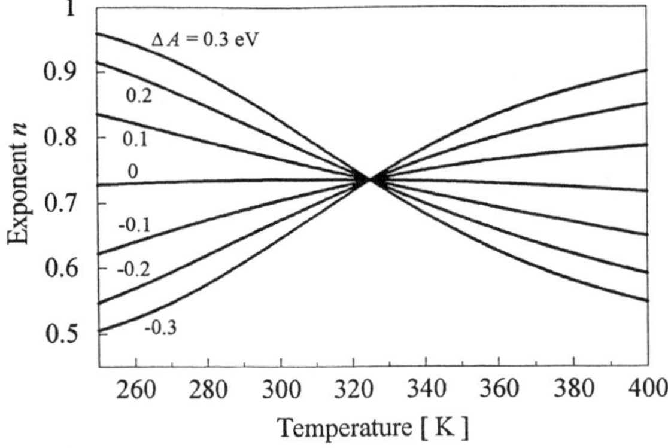

Fig. 4.23 Transition from chemisorption control to diffusion control by temperature variation in simulated experiments for various differences in the activation energies $\Delta A = A_{ph,ch} - A_{diff,\beta}$

4.4.4 Conclusions

The model calculations demonstrate that the absorption kinetics of a system can lie in a transition region where not one but two or more partial steps are rate determining. In this case, it can easily happen that the range of accessible experimental parameter variations does not suffice to determine the features of a reaction mechanism completely with inclusion of all limits. A change of three to four orders of magnitude can be required to describe the transition region for such a mixed mechanism correctly. Therefore, one has to be careful if activation energies derived from extrapolations in transition regions are taken as quantitative data. The findings of the present model calculations that the transition between two limiting regions of reaction kinetics can be stretched over several orders of parameter variations is quite a normal behavior for reactions composed of a sequence of partial steps. In Sect. 6.2 a poisoning model is discussed of a system coated with an impurity layer of constant thickness. The curves describing the kinetics in the transition region look very similar to those presented here. In models on the kinetics of absorption and desorption reactions the approximation of only one rate determining step is only then justified if all other partial fluxes are at least one order of magnitude larger than the slowest one over the whole range of an experiment. This result sounds quite reasonable and should be kept in mind when experimental data are compared with the relative simple laws derived for well-defined limiting cases.

4.4.5 Desorption

Structure of the model
In the advanced model for hydrogen absorption discussed in the preceding sections, the rate laws for the forward and backward reaction at each of the partial steps are already included. If the hydrogen pressure in the receiver has values below the decomposition pressure of the hydride and the initial state of the sample is a hydride a desorption run is simulated automatically. However, the size of the program and the run time of the numeric calculations is reduced if separate models are used for the absorption and the desorption process where all routines not

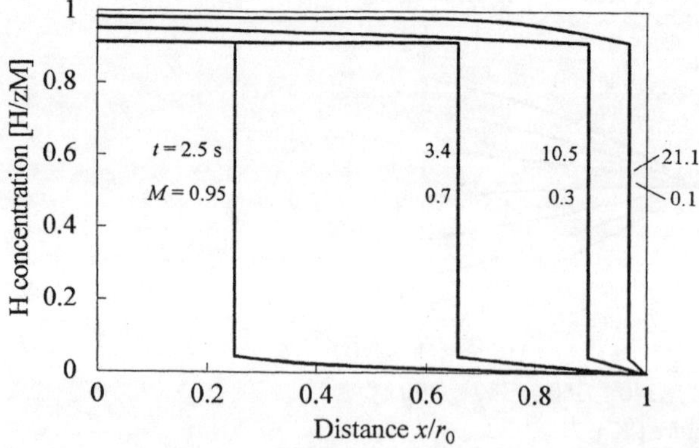

Fig. 4.24 Concentration profiles in a spheric sample during dehydriding. The hydride formed is H_2M, and the symbol M represents the relative hydrogen concentration $\bar{c}/\bar{c}_{max}$

needed are canceled. Each simulation run provides, again, a concentration profile for a given reaction time or for the normalized hydrogen quantity, M, removed from a spherical sample. An example for this kind of simulations is shown in Fig. 4.24. The shell of α solid solution on the sample surface is increased with increasing removal of hydrogen. Most of the hydrogen gas released is created by decomposition of the β hydride at the α/β interface. Concentration changes in the β phase contribute little to the hydrogen flux. This has the consequence that hydrogen diffusion in the β phase must not be considered as a rate determining step. Only hydride decomposition and diffusion in the α phase, surface penetration or recombination of chemisorbed hydrogen atoms to hydrogen molecules can act as a bottle neck for the reaction rate.

Transition from Recombination to α-Phase Diffusion Control
The curves presented in Figs. 4.25, 26 are simulations for decomposition of a spheric hydride sample in a receiver with very low H_2 pressure. Thus, it considers only the desorption steps in the overall process. The log M versus log t plot in Fig. 4.25 displays four curves for different values of the activation barrier for chemisorption, $A_{ph,ch}$ (Fig. 4.13). Since the activation energy for the recombination of chemisorbed hydrogen atoms, $A_{ch,ph}$, is composed of the chemisorption energy, ΔH_{chem}, and the activation energy for chemisorption, $A_{ph,ch}$, it is increased, if $A_{ph,ch}$ is increased. For high activation energy values the desorption reaction is surface controlled and $v \propto t$ holds. The concentration of the α phase remains constant and has the high value of the α/β equilibrium up to the very end of the run. Therefore, the linear rate law is maintained in this model until hydrogen is almost totally removed from the sample.

If $A_{ph,ch}$ is small, the reaction becomes, at least partially, controlled by diffusion in the α phase. In this limit $v \propto \sqrt{t}$ is approached and the reaction rate is decreased due to the reduction of the concentration gradient with increasing α layer thickness. Figure 4.26 shows the change of the fluxes for hydride decomposition, for α phase diffusion and for recombination of hydrogen atoms on the sample surface for three different values of $A_{ph,ch}$ as a function of the amount of removed

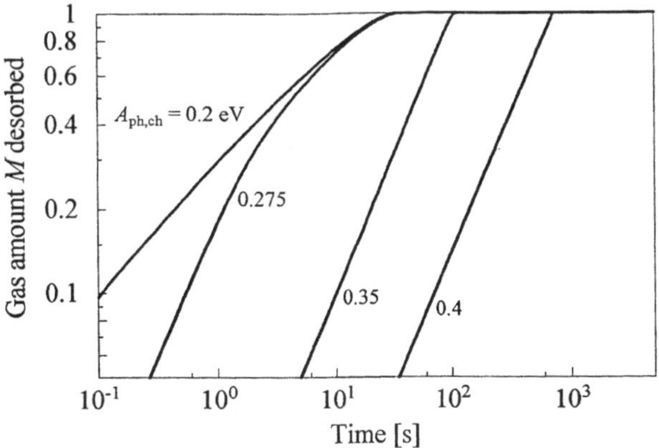

Fig. 4.25 Simulated hydrogen desorption curves for various values of the chemisorption activation energy $A_{ph,ch}$

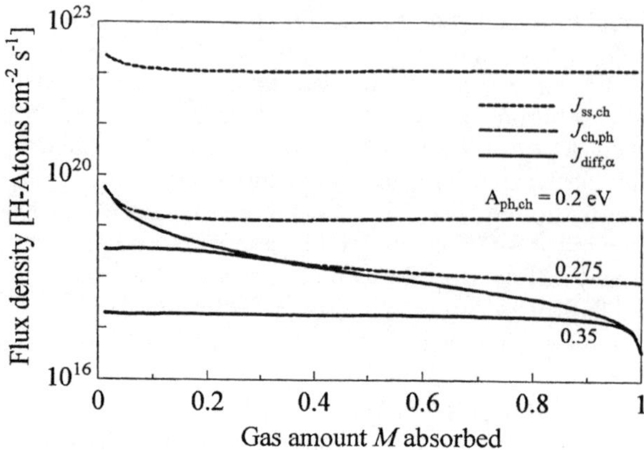

Fig. 4.26 Hydrogen flux densities of partial fluxes for chemisorption, $J_{ch,ph}$, surface penetration, $J_{ss,ch}$, and diffusion, $J_{diff,\alpha}$ for desorption runs of Fig.4.25

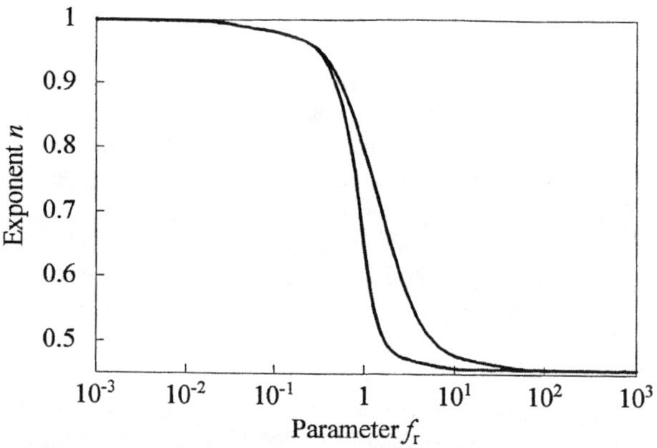

Fig. 4.27 Transition from recombination to diffusion control during desorption as a function of the flux ratio $f_r = J_{ch,ph}/J_{diff,max}$

hydrogen, M. The rate controlling step is the smallest one of these fluxes. For $A_{ph,ch} = 0.35$ this is the recombination partial flux, almost to the very end of the run. For $A_{ph,ch} = 0.2$ eV it is, after a short initial period of time, hydrogen diffusion in the α phase and for $A_{ph,ch} = 0.275$ eV it is for M < 0.3 recombination control and for M > 0.5 diffusion control.

In Fig. 4.27 the transiton from recombination control to diffusion control is demonstrated in a plot exponent n in the time law $M \propto t^n$ versus the ratio f_r of the partial fluxes for recombination, $J_{ch,ph}$, and the maximum diffusion flux in the α phase, $J_{diff,max}$. Again, the lower curve gives the result for the exact simulation and the upper one estimated data obtained from the limits for models with only one rate determining partial step. The transition here from one rate determining step to the other is faster than in the examples for hydrogen absorption. There are several reasons for this different behavior. One of it may be the close coupling between

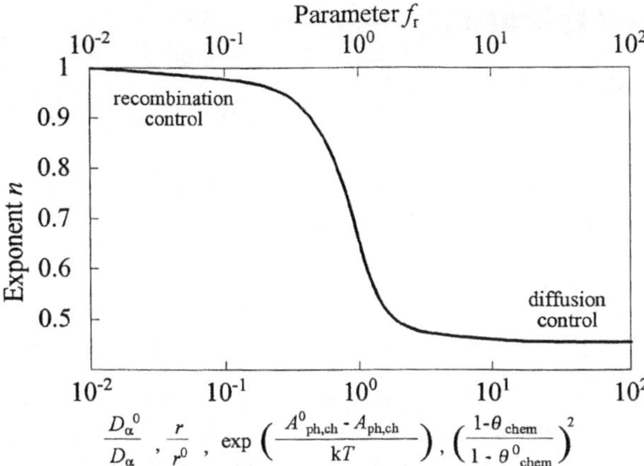

$$\frac{D_\alpha^0}{D_\alpha} \; , \; \frac{r}{r^0} \; , \; \exp\left(\frac{A^0_{ph,ch} - A_{ph,ch}}{kT}\right) , \left(\frac{1-\theta_{chem}}{1-\theta^0_{chem}}\right)^2$$

Fig.4.28 Exponent n of the time law $M \propto t^n$ as a function of system parameters, normalized to a value where n = 0.75 holds, for the desorption reaction

diffusion flux and the concentration of chemisorbed atoms. As long as the diffusion flux is fast $\theta_{H(chem)}$ remains constant at a relatively high level. If diffusion becomes slower $\theta_{H(chem)}$ decreases rapidly together with the concentration at the subsurface layer in the metal. In addition, the rate of the recombination reaction decreases proportionally to the square of the concentration of chemisorbed atoms, $\theta_{H(chem)}$. In the absorption model the situation is quite different. There the chemisorption reaction is only affected by the availability of vacant surface sites.

In Fig. 4.28 the lower curve from Fig. 4.27 is replotted as a function of normalized parameters. The meaning and interpretation of the curves are analogous to the examples for absorption discussed in Sects. 4.4.3, 4 and may easily be duplicated by the reader.

5 Low-Temperature Oxidation

The normal surface state for metallic materials is an oxide skin since a bare surface would be a highly unstable chemical situation in the oxidizing air atmosphere of our earth. Most metals form at ambient and even at low temperatures thin adherent oxide scales which protect the bulk of the metal from further attack by reactive gas molecules. The growth of such tarnishing layers is a complex process. The reacting elements, metal atoms and oxygen molecules, are separated by the oxide skin formed as reaction product. This is an ionic compound and, therefore, the more mobile atom of the two reactants has to be transferred first to a charged defect in the crystal lattice before it can migrate to the reaction front. An electric field can exist across the oxide layer and introduce additional effects that make understanding of the reaction mechanism more difficult.

The first experimental and theoretical investigations on the kinetics of metal oxidation yielded the $l \propto \sqrt{t}$ law found by *Tammann* [5.1] and *Pilling* and *Bedworth* [5.2]. They showed that this time law is obtained if diffusion of atomic species through a growing oxide scale is the rate determining step. The mechanism of high-temperature oxidation is well described by the theory of *Wagner* [5.3, 4]. The basic conception of his oxidation models is the quantitative description of the oxidation mechanism by linear transport equations for charged defects through an ionic crystal and not by diffusion of neutral particles. In his treatment he could take advantage of the already existing theory of ion transport in electrolytes and he has demonstrated that close relations exist between oxidation rates and the electric conductivity of oxides. *Wagner* obtained a parabolic time law $l \propto \sqrt{t}$ for oxide layer growth since diffusion is assumed to be the rate determining step. This assumption is justified for thick oxide layers where processes at interfaces can be approximated by the equilibrium conditions. The constant of proportionality, the so-called scaling constant, can be expressed as a function of microscopic tracer diffusion coefficients of migrating defects. *Wagner* was the first who focused attention to the significance of overall charge neutrality in the kinetics of dry oxidation. If a stoichiometric oxide is formed and anions, cations, and electrons participate in the reaction, then the total charge transported through the oxide scale under steady-state conditions must be zero at any site and at any time. He developed the conception of the so-called ambipolar diffusion which is equivalent to the coupled-currents principle used in this work. Further developments of Wagner's theory by *Hauffe* [5.5] and other researchers discussed the role of space and surface charges which affect the expressions for scaling constants. A great many experimental data proving the theory have been published in an abundant literature on oxidation kinetics in the last decades.

In high-temperature oxidation the transport of the reactants, oxygen and metal, through the oxide layer is mainly accomplished by thermally activated diffusion. This mechanism is, however, unable to explain the very fast initial oxidation process observed at low temperatures. *Cabrera* and *Mott* [5.6] were the first to point out that the low ion mobility at low temperatures can be enhanced strongly by a contact potential produced by electron transfer from the metal to the oxide surface. According to the assumption that electronic equilibrium exists they derived a parabolic and an inverse logarithmic rate law for the initial stage of oxidation. These limits will be mentioned again below.

In contrast to the single-current approach of *Cabrera* and *Mott* in the models on low-temperature oxidation developed by *Fromhold* [5.7] the conception of coupled currents has been consequently applied. He considers at least one ionic and one type of electronic currents. This procedure resembles to Wagner's ambipolar diffusion treatment. Depending on the mechanism assumed for the electron transport *Fromhold* has simulated many types of oxide growth curves and contact potential profiles for the limit of constant defect concentrations at the interfaces. But in his treatments two aspects are still missing: Inclusion of the concentration changes caused by interface reactions and the discussion of the effect of different types of lattice defects on the transport mechanism and on the time laws. Pressure and temperature effects observed in experimental oxidation curves cannot be explained properly by reaction mechanisms neglecting surface processes. Also, his comprehensive models are not yet fully consistent simulations of real oxidation curves. In the more advanced reaction models developed [5.8–11] the reactions at both interfaces of the oxide scale are incorporated as additional partial steps of the mechanisms. Thus, they enable simulation of oxidation processes with dispense of the unrealistic assumptions which had been introduced by the authors of preceding treatises on metal oxidation to simplify the mathematical treatment. The models on low-temperature oxidation kinetics presented in this chapter of the book discuss only reaction mechanisms which have this complete and consistent structure.

5.1 Experimental Results

Typical experimental curves on metal oxidation at low and elevated temperature are presented in the book of *Fehlner* [5.12]. Here, the general feature of oxidation curves is explained by some results for metal film samples measured with a quartz crystal microbalance in isothermal-isobaric experiments and other data measured with the volumetric Wagener method. The examples depicted show different pressure and temperature effects. A system with almost pressure-independent absorption kinetics and another with a very strong pressure dependence is presented. Other results show the temperature dependence of the oxidation process and the large difference in layer thickness observed with individual metals. The reaction models discussed in subsequent sections should be able to simulate, at least qualitatively, the feature of the various types of experimental oxidation curves.

Oxidation in the system vanadium-oxygen dependents strongly pressure (Fig. 5.1) but this effect is much less pronounced in the system titanium-oxygen (Fig. 5.2). At temperatures below 50 °C the layer thickness of the oxide scale on aluminum is about one monolayer (Fig. 5.3) whereas the oxide skin on iron film samples is about ten times thicker (Fig. 5.4).

The oxidation curves begin with an initial absorption process which is too fast for quantitative experimental determination. The rate of the subsequent oxidation

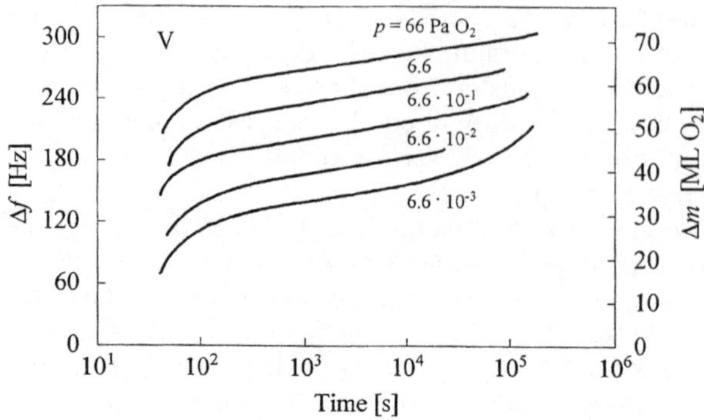

Fig. 5.1 Oxide growth curves of vanadium films measured at 35 °C and different pressures [5.13]

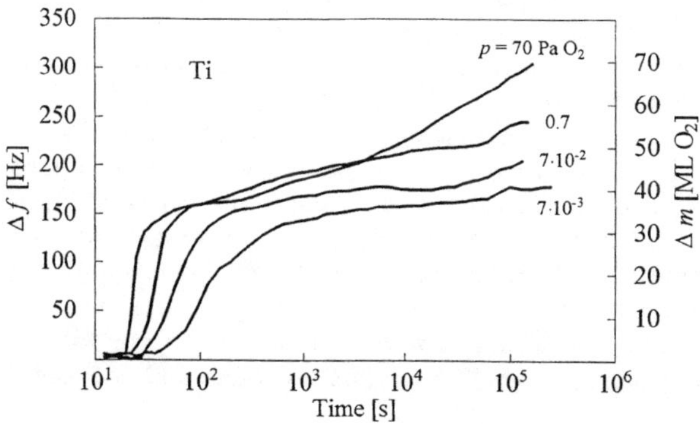

Fig. 5.2   Oxide growth curves of titanium films measured at 50 °C and different oxygen pressures [5.14]

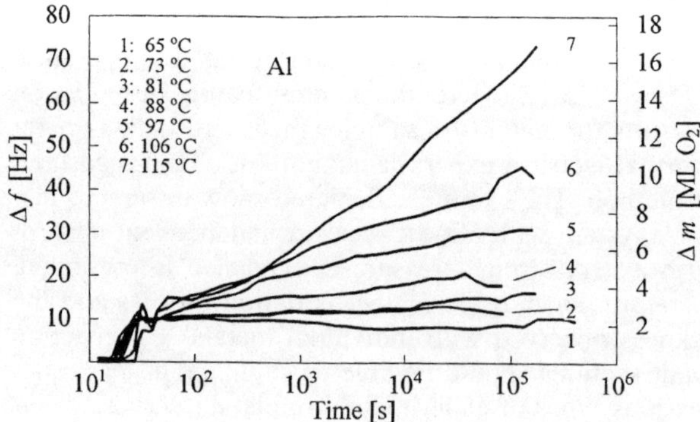

Fig. 5.3 Oxide growth curves of aluminum films measured at 0.7 Pa O_2 and different temperatures [5.14]

region is very slow at low temperature. At more elevated temperatures the reaction rate is increased and the oxide layers are becoming relatively thick after some hours of exposure (Fig. 5.4). Increasing of the exposure pressure in steps of one hour during oxidation of a vanadium film does not shift the absorption curve from

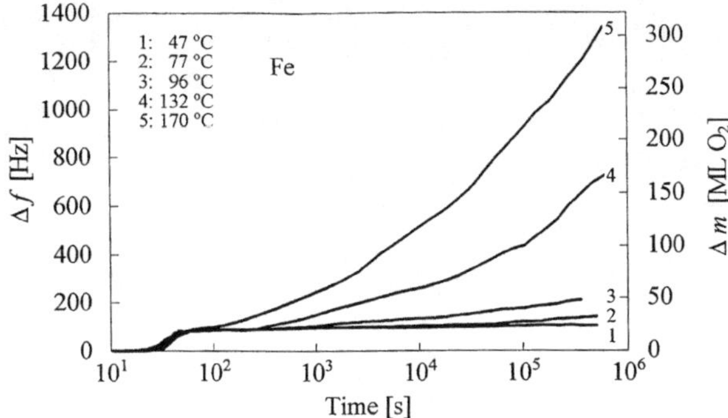

Fig. 5.4 Oxide growth curves of iron films measured at 0,7 Pa O_2 and different temperatures [5.14]

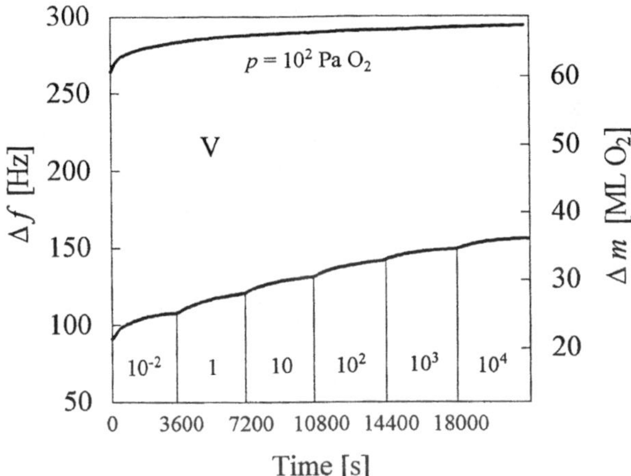

Fig. 5.5 Curves weight gain versus time of two vanadium films exposed to oxygen at 35 °C; (a)at a constant, relative high, pressure, and (b) with stepwise pressure increases [5.13]

one isobar in Fig. 5.1 to the next one with much higher thickness but yields only a very small additional thickness growth (Fig 5.5). A similar experiment with pressure changes in the course of two exposures of a titanium film to oxygen shows that the reaction rates can be changed by such treatments but it is not simply a transition from one isobaric oxidation curve to another (Figs. 5.2, 6). Further results of studies on the low-temperature oxidation of metal films measured with a quartz crystal microblance are published in Refs. [5.13, 14].

The oxidation curves measured with initially gas-free surfaces can be subdivided into three characteristic stages: A very fast initial stage, a slower logarithmic second stage and finally a third stage with a time law faster than logarithmic.

Additional information on the very fast initial section of the oxidation reaction can be obtained from curves reaction probability r vs. amount of absorbed oxygen, N, measured with the volumetric method at very low exposure pressures. The curves in Fig. 5.7 show that the initial reaction probability at low pressures has the maximum value of one. The amount of oxygen absorbed before the reaction

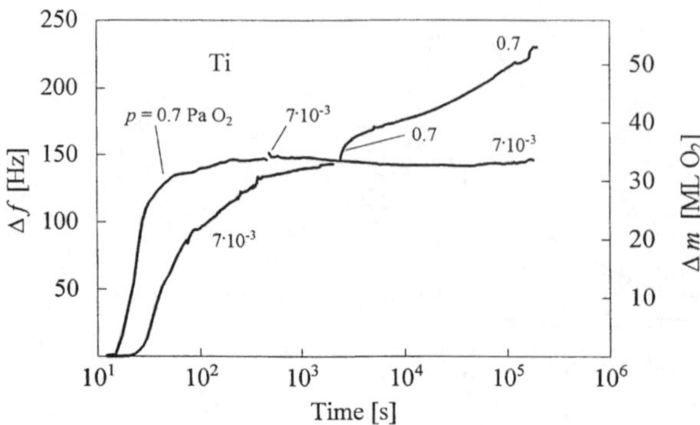

Fig. 5.6 Two exposure runs of titanium films at 60 °C with pressure changes at the times indicated on the curves [5.14]

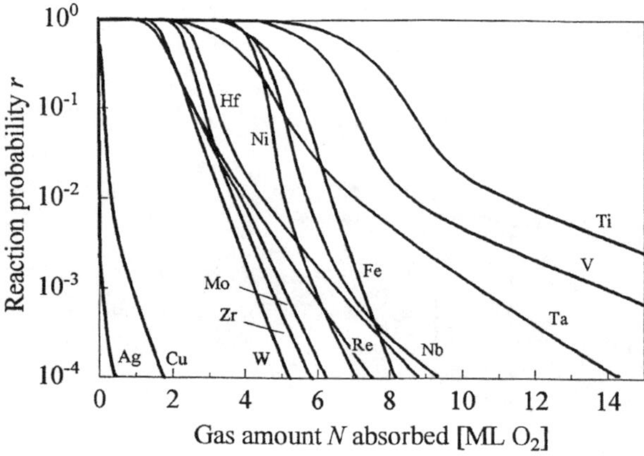

Fig. 5.7 Reaction probability r of oxygen with various metal film samples at 300 K as a function of the amount of absorbed oxygen [5.15]

probability decreases exponentially with increasing N depends, again, on the metal investigated. It corresponds to the hight of the fast absorption step in isobaric-iso-thermal experiments (Figs. 5.3, 4). Due to the roughness of film surfaces the amount of oxygen absorbed per unit of geometric sample surface area is by a factor two to five larger than for metal sheet samples. In spite of this systematic error in the absolute determination of the thickness of oxide layers measured with film samples, there is no doubt that the thickness varies from about one monolayer for aluminum to ten or more layers found with many transition metals [5.15].

5.2 Rate Laws Proposed in the Literature

Various empirical rate laws have been proposed for oxidation reactions at low-temperatures: parabolic, linear, sigmoidal and logarithmic [5.16]. They are suited for the presentation of experimental data but frequently they cannot provide much assistance for the interpretation of the reaction mechanism. Curve fitting is sometimes possible in the same system by different rate laws depending on the temperature range or the oxide layer thicknesses chosen in an experiment. The kinetics then can be formally attributed to different reaction mechanisms. This is especially true if experimental data are measured in the transition range between two branches of a complex oxidation mechanism or if the initial coverage of the sample surface is not defined correctly. Rate equations for simple models are derived using the approximation that only one of the partial processes involved dominates the process, for example, one of the defect currents. The faster currents are then assumed to be in virtual equilibrium while the slowest one is the rate determining step. These are so-called single current approaches. More advanced treatments consider several processes and analyze the interactions between them. This conception is the central point of the coupled current treatments applied by *Fromhold* [5.7] and others.

5.2.1 Parabolic Law

The basis of the parabolic law is the linear transport equation

$$J = -D\nabla c(l, t) + z_i b E c(l, t), \tag{5.1}$$

where b is the mobility and z_i the charge of the diffusing species i. This equation is valid for neutral as well as for charged species in low electric fields and yields after integration of the oxide growth equation $d[l(t)]/dt \propto J$ the well-known parabolic rate law $l \propto \sqrt{t}$. Diffusion control is the most commonly assumed mechanism in early theories on various limiting cases as discussed, for example, by *Hauffe* [5.5] and *Evans* [5.16]. *Cabrera* and *Mott* [5.6] derived a parabolic growth law for very thin layers under the condition that the migration of cations is enhanced by an intrinsic electric field across the oxide layer which is produced by tunneling electrons.

5.2.2 Inverse Logarithmic Law

Cabrera and *Mott* [5.6] also derived a rate law for the special case that incorporation of metal cations into interstitial sites of the oxide is rate determining. The oxide growth rate is then proportional to the probability of surmounting the potential barrier between the sites of the atoms in the metal and interstitial sites in the oxide. The potential barrier is assumed to be potential dependent and integration of the rate law yields the Cabrera-Mott inverse logarithmic time law $l_0/l = \text{const} - \log t$.

5.2.3 Linear Law

If the oxygen pressure is very low, the oxygen supply can become rate determining and the oxide growth rate is no longer dependent on the oxide layer thickness. According to the laws of kinetic gas theory the reaction rate is proportional to the pressure p_{O_2}. Then $d[l(t)]/dt \propto p$ holds or $l(t) \propto pt$.

5.2.4 Logarithmic Law

Electronic rate control of the oxidation process at low temperatures was discussed first by *Mott* [5.17]. For very thin oxide layers where electron tunneling prevails the electronic current depends on the tunneling probability which is proportional to $\exp(-\operatorname{const} l)$. Thus, the oxide growth rate decreases exponentially with increasing layer thickness and a logarithmic time law $l \propto \log t$ is obtained by integration.

5.3 Partial Steps of the Oxidation Reaction

5.3.1 Reaction Mechanisms

If adherent oxide scales are formed on a metal surface the reaction

$$x\,M + y/2\ O_2 = M_xO_y$$

cannot proceed as written in the formula since the reacting elements metal and oxygen are separated by the oxide film formed after a very short initial period of time. Therefore, the reaction mechanism must be separated into several partial steps which include generation, transport, and annihilation of intermediate reaction products. Depending on the type of migrating lattice defects the reaction front can be either at the oxide/gas or at the oxide/metal interface (Fig. 5.8). Atomic or ionic metal or oxygen species are transported as interstitials or as vacancies through the oxide scale and an electronic current from the metal phase to the oxide

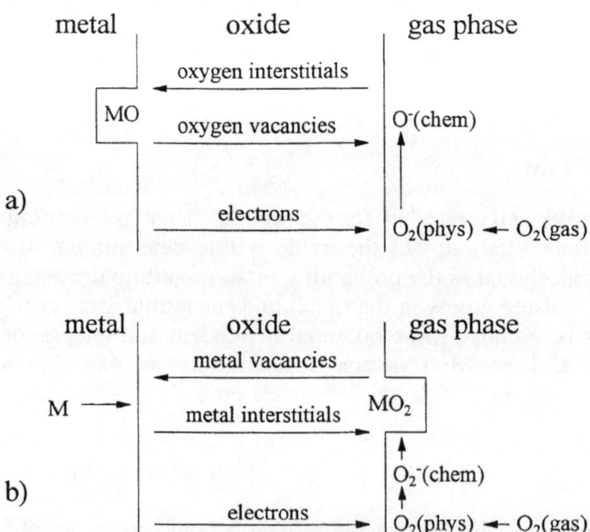

Fig. 5.8 Partial steps of the oxidation reaction. a) transport of oxygen ions or vacancies; reaction front is the metal/oxide interface, and (b) transport of the metal ions or vacancies; reaction front is the oxide surface

surface is needed to maintain overall charge neutrality. In addition reactions between lattice defects, electrons and adsorbed species occur at both interfaces.

The reaction between gaseous and adsorbed oxygen at the oxide surface proceeds in two steps. At first, oxygen molecules are weakly bonded to the oxide surface or physisorbed. The physisorbed oxygen species are acceptors for electrons from the metal phase and strongly bonded chemisorption particles can be formed such as O_2^-, O^{2-} or O^-. Various surface and bulk defects are discussed in the literature and in the appendix. They provide a large number of candidates for reactants of partial reactions that can be incorporated in a model of the reaction mechanism. To restrict the number of free parameters the mechanisms should contain, if possible, only one chemisorption, one ionic, and one electronic species. This simplification is justified for the simulation of real systems if in the pressure and temperature range of interest the concentration of neglected species is small and does not contribute much to the ionic and electronic fluxes.

The rate equations of all interface and surface reactions and of the transport mechanisms of the mobile species provide a set of mathematical expressions which enable the calculation of all currents and concentrations of the defects considered by a similar numerical treatment as applied for the hydriding problem in the preceding chapter. Finally, the oxidation curves given as mass gain versus time can be obtained by integration of the ion current.

For bulk defects the symbols proposed by *Schottky* [5.18] are used in the models in a slightly modified style (Appendix A). They are for metal and oxygen interstitials M^{z+} and O^{z-} for vacancies $[M]^{z-}$ and $[O]^{z+}$ with z valency of the positive and negative charge of the defects considered. The meaning of the defect symbols has been discussed in detail by *Rickert* [5.19]. The different states for metal and oxygen atoms occurring in the models, such as molecular, atomic, ionic, physisorbed or chemisorbed, and the relations needed in models on oxidation processes are discussed in Appendix A and in Chaps. 2, 4. It is anticipated in this chapter that the reader is already familiar with the fundamentals of reaction kinetics. Therefore, here such phenomena will be mainly investigated which differ from the models of hydride formation.

5.3.2 Charge Distribution and Electric Fields

The presence of charged defects in the system metal/oxide/oxygen causes electric fields. If the model system is assumed to be planar and only one ionic defect species is considered then the charges are located for interstitials on planes between two regular atom layers in the oxide or for vacancies in the atom planes itself. This convention for the description of charges and the electric field in the models is displayed in Fig. 5.9. The planar system is subdivided into L parallel planes with the defect concentrations $\theta_e(0)$, $\theta_i(l)$, ..., $\theta_i(L + 1)$. Electronic excess charges $\theta_e(0)$ are assumed to exist only at the metal phase boundary. The quantity a denotes the distance between the lattice planes. The potential barrier with the height A corresponds to the barrier between two adjacent equilibrium positions of defects in the oxide. The plane 0 is the uppermost metal layer and this is also the position of the electronic compensation charge $\theta_e(0)$. The plane L represents the oxide surface. The space between adjacent charge planes is assumed in the models to be free of charges and exhibit a homogeneous field strength $E(0)$, ..., $E(L + 1)$.

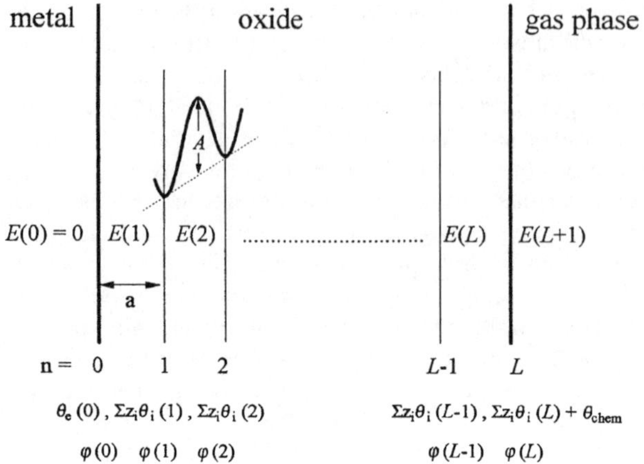

Fig. 5.9 Charge and electric field distribution in the model system metal/oxide/oxygen. The planes 0 to L are equilibrium positions of defects

The corresponding potentials are $\varphi(0), \ldots, \varphi(L)$. θ_{chem} is the surface charge density given by the sum of the charges of chemisorbed species per unit area.

According to Gauss' law of electrostatics [5.20] the difference between the values of the electric field on two sides of a charged plane is proportional to the areal density of the charge caused by n defects at that plane:

$$\sum_{i=1}^{j} z_i \cdot \theta_i(n) = \epsilon_0 \epsilon_{ox}[E(n+1) - E(n)], \quad n = 1, \ldots, L-1. \tag{5.1}$$

Inside the metal the field becomes zero, $E(0) = 0$, and the electronic charge density at position 0 is given by

$$\theta_e(0) = \epsilon_0 \epsilon_{ox} E(1) \tag{5.2}$$

On the oxide surface the following relation holds:

$$\sum_{i=1}^{j} z_i \cdot \theta_i(L) + \theta_{chem} = \epsilon_0 \epsilon_{O_2} E(L+1) - \epsilon_0 \epsilon_{ox} E(L). \tag{5.3}$$

In dry oxidation no excess charges exist in the system and the condition of overall charge neutrality yields the equation

$$\theta_e(0) + \sum_{i=1}^{j} \sum_{n=1}^{L-1} z_i \cdot \theta_i(n) + \sum_{i=1}^{j} z_i \cdot \theta_i(L) + \theta_{chem} = 0. \tag{5.4}$$

The first term represents the electronic charge in the metal, the second the defect charges in the bulk of the oxide and the last two terms the surface charges. This has the consequence that a negative defect charge in the oxide and/or on the oxide surface must be compensated by a positive charge in the metal. If the concentration terms in (5.4) are substituted by the field terms (5.1–5.3) the relations

$$\epsilon_0\epsilon_{ox}\left(E(1) + \sum_{n=1}^{L-1}[E(n+1) - E(n) - E(L)]\right) + \epsilon_0\epsilon_{O_2}E(L+1) = 0$$

or

$$\epsilon_0\epsilon_{ox}[E(1) + E(L) - E(1) - E(L)] + \epsilon_0\epsilon_{O_2}E(L+1) = 0$$

are obtained which demonstrate that $E(L+1) = 0$ holds. Thus, also in the gas phase at position $L + 1$ the field becomes zero in any case.

5.3.3 Reactions at the Metal/Oxide Interface

At the metal/oxide interface two different processes must be distinguished. If the reaction front is there, the oxide molecules can be formed either by generation of oxygen vacancies $[O]^{z+}$

$$xM \leftrightarrows M_xO_y + y[O]^{z+}\,(oxide) + y\,z\,e\,(metal) \tag{5.5}$$

or by annihilation of O^{z-} interstitials:

$$x\,M + y\,O^{z-}(oxide) \leftrightarrows M_xO_y + y\,z\,e\,(metal). \tag{5.6}$$

If the metal/oxide interface is not the reaction front either metal interstitials M^{z+} are created there

$$M \leftrightarrows M^{z+}(oxide) + z\,e\,(metal) \tag{5.7}$$

or $[M]^{z-}$ vacancies are annihilated

$$M + [M]^{z-} \leftrightarrows z\,e\,(metal). \tag{5.8}$$

It is important to note that in all four cases electrons are produced in the metal phase which have to be transported to the oxide surface.

5.3.4 Reactions at the Oxide Surface

The situation at this interface is more complex. Oxygen must be at first fixed to the oxide surface by the processes of physisorption and chemisorption:

$$physisorption:\ O_2(gas) \leftrightarrows O_2(phys); \tag{5.9}$$

$$molecular\ chemisorption:\ O_2(phys) + ze \leftrightarrows O_2^{z-}(chem); \tag{5.10}$$

$$dissociative\ chemisorption:\ O_2(phys) + 2ze \leftrightarrows 2O^{z-}(chem). \tag{5.11}$$

The adsorbed oxygen species can then undergo reactions with other defects on the oxide surface. If the oxide surface is the reaction front, oxide can be formed by annihilation of M^{z+} interstitials

$$xM^{z+}(\text{surface}) + \left(\frac{zx}{\beta}\right)O_\delta^{\beta-}(\text{chem}) \leftrightarrows M_xO_y + \left(\frac{zx\delta}{2\beta} - \frac{y}{2}\right)O_2(\text{phys}) \qquad (5.12)$$

with $\delta = 1$ for dissociative chemisorption and $\delta = 2$ for molecular chemisorption. But it can also be produced by generation of $[M]^{z-}$ vacancies

$$\left(\frac{zx}{\beta}\right)O_\delta^{\beta-}(\text{chem}) \leftrightarrows x[M]^{z-} + M_xO_y + \left(\frac{zx\delta}{2\beta} - \frac{y}{2}\right)O_2(\text{phys}). \qquad (5.13)$$

If the oxide surface is not the reaction front then O^{z-} interstitials are formed

$$O_\delta^{\beta-}(\text{chem}) \leftrightarrows (\beta/z)O^{z-}(\text{surface}) + \left(\frac{\delta}{2} - \frac{\beta}{2z}\right)O_2(\text{phys}) \qquad (5.14)$$

or $[O]^{z+}$ vacancies can be annihilated

$$(\beta/2)[O]^{z+}(\text{surface}) + O_\delta^{\beta-}(\text{chem}) \leftrightarrows \left(\frac{\delta}{2} - \frac{\beta}{2z}\right)O_2(\text{phys}). \qquad (5.15)$$

5.4 Relations and Constants Used in Model Calculations

This section presents the structure of models which simulate the formation of semiconducting coatings on metals and points to approximations used in mathematical treatments.

5.4.1 Equation of Continuity

The equation of continuity relation guarantees the conservation of matter at each point of the system and must be obeyed strictly. It can be written as

$$\frac{\partial c_i(x, y, z, t)}{\partial t} + div\, J_i(x, y, z, t) + \sum_{\text{reactions},r} (v_{r+} - v_{r-}) = 0, \qquad (5.16)$$

where $c_i(x, y, z, t)$ is the density of the defect i in m^{-3}. The reduction of the model to an one-dimensional system simplifies the problem and (5.16) can be written as

$$\frac{\partial c_i(x, t)}{\partial t} - \frac{\partial J_i(x, t)}{\partial x} + \sum_{\text{reactions},r} (v_{r+} - v_{r-}) = 0. \qquad (5.17)$$

The reaction rates v_{r+} and v_{r-} give the increase or decrease of the number of particles i in units $m^{-3}s^{-1}$ caused by the reaction r.

5.4.2 Steady-State Condition

Most mathematical treatments on reaction kinetics introduce the approximation that after a short incubation period the concentration changes $\partial c/\partial t$ in the system are small when compared with the divergence of the current density and the reaction rates. Then the first term in (5.17) can be canceled. This rather realistic approach is termed steady state and (5.17) can be written as

$$-\frac{\partial J_i(x, t)}{\partial x} + \sum_{\text{reactions},r} (v_{r+} - v_{r-}) = 0. \tag{5.18}$$

In the bulk of the oxide no reactions shall occur and according to (5.18) the ion current must be constant:

$$\frac{\partial J_i(x, t)}{\partial x} = 0 \quad or \quad J_i = \text{constant}. \tag{5.19}$$

For a model configuration (Fig. 5.9) this has the consequence that the difference between the forward and reverse ion current must be constant for all planes occupied by defects.

$$J_{i+} = J_{i-} = j_{ion} = \text{constant}. \tag{5.20}$$

At phase boundaries where chemical reactions take place integration of (5.18) between planes 0 and 1 or L-1 and L, respectively (Fig 5.9), yields

$$J_{i+} - J_{i-} + \sum_{\text{reactions},r} (v_{r+} - v_{r-}) = 0. \tag{5.21}$$

The reaction rates v_{r+} and v_{r-} must be given in the same units as the currents, in ML/s or $m^{-2}s^{-1}$.

5.4.3 Principle of Coupled Currents

If the mobile defects are ions in addition to matter then charge is also transported through the oxide layer. The charge flux is characterized by the effective charge z_i of a defect and the direction of the particle current. A current directed from the metal phase to the oxide surface is defined here as positive. The electric field in the system is affected by the currents of charged defects. However, in steady state the charge transfer of the individual currents must compensate to zero since otherwise extremely high electric field would be produced. This condition requires that no net charge flux exists

$$\sum_{i=0}^{j} z_i J_i = 0. \tag{5.22}$$

This principle of coupled currents is a very useful approximation and it is exactly obeyed for permeation fluxes through a layer which does not grow. The error introduced by the approximation of the coupled currents for growing films is small and can be estimated. If the thickness of an oxide layer is increased from l to $l + a$

while the potential V across the oxide film is kept constant, the surface charge density is given by

$$\theta(l) = \epsilon_0 \epsilon_{ox} V / l \tag{5.23}$$

and

$$\theta(l + a) = \epsilon_0 \epsilon_{ox} V / (l + a). \tag{5.24}$$

The difference between both concentrations is

$$\Delta\theta = \theta(l) - \theta(l + a) = \epsilon_0 \epsilon_{ox} V \left(\frac{1}{l} - \frac{1}{l+a} \right). \tag{5.25}$$

$\Delta\theta$ corresponds to a small net current which violates the coupled currents principle.

$$\Delta J = J_{el} - J_{ion}, \tag{5.26}$$

$$\Delta\theta = \epsilon_0 \epsilon_{ox} V \left(\frac{1}{l} - \frac{1}{l+a} \right) = \Delta J \, \Delta t, \tag{5.27}$$

where Δt is the growth time for one atomic layer $\Delta l = 1$. On the other hand, the oxide growth rate is proportional to the ion current:

$$\frac{d[l(t)]}{dt} = R J_{ion} \tag{5.28}$$

with the constant of proportionality, R. For finite differences and Δt one can write:

$$\frac{\Delta l}{\Delta t} = R J_{ion} \quad \text{with} \quad \Delta l = a \quad (a = 3 \cdot 10^{-10}\,\text{m}). \tag{5.29}$$

Inserting Δt from (5.29) into (5.27) the relative error $\Delta J / J_{ion}$ caused by the principle of the coupled currents is given by

$$\frac{\Delta J}{J_{ion}} = R \epsilon_0 \epsilon_{ox} V \left(\frac{1}{l} - \frac{1}{l+a} \right). \tag{5.30}$$

For typical values of $R = 1.3$, $\epsilon_{ox} = 13$ and $V = 0.5$ V one obtains

$$
\begin{aligned}
l &= 5a, & \Delta J / J_{ion} &= 10^{-2}; \\
l &= 10a, & \Delta J / J_{ion} &= 3 \cdot 10^{-2}; \\
l &= 50a, & \Delta J / J_{ion} &= 10^{-4}.
\end{aligned}
$$

Thus, the error for a layer thickness $d > 5a$ ($a = 3 \cdot 10^{-10}$ m) is less than 1%. This demonstrates that the principle of coupled currents is a powerful and reliable approximation.

5.4.4 Structure of the Models

The number of equations required for the calculation of rate and time laws depends on the number of defects and reactions incorporated in a specific model system. Because of the large number of possible reactions and defects, it would be unreasonable to define many elements for one big comprehensive model with many intermediate species. Developing individual models with a minimum number of parameters for simple systems of specific interest is much more economical. The four most important point defect types in an oxide lattice are metal interstitials, M^{z+}, metal vacancies, $[M]^{z-}$, oxygen interstitials, O^{z-}, and oxygen vacancies $[O]^{z+}$. They suggest a natural choice for four basic model structures. For sake of simplicity, from the various types of chemisorption species $O_\delta^{\beta-}$ such as, O_2^-, O^-, or O_2^- only one chemisorption particle should be chosen with fixed indices β and δ. The choice of a specific defect acting as the most mobile species in the bulk of the oxide defines also the reaction equations for formation and decomposition. Relevant equilibrium constants and rate laws must be established and linked according to the conditions of charge neutrality and coupled currents.

5.4.5 Numerical Procedures

The numbers of unknown variables and of equations available for numerical solutions for a specific system are listed in Tables 5.1, 2. Both numbers are equal and, consequently, the problem has exactly one solution. The set of nonlinear equations of the variables of interest can be solved numerically. The experimental para-

Table 5.1 List of unknown variables

variable	number
concentrations: $\theta_i(n)$; $i = 1\dots j$, $n = 1\dots L$	$j \cdot L$
electron concentration: $\theta(0)$	1
electronic and ionic currents: J_i	j
electric field strength: $E(n)$; $n = 0\dots L+1$	$L+2$
Total	$jL + j + L + 3$

Table 5.2 List of available equations

equations available	number
steps of the electric field strength : $\displaystyle\sum_{i=1}^{j} z_i\theta_i(n) = \epsilon_0\epsilon_{ox}(E(n+1) - E(n));\quad n,\dots,L$	$L+1$
interface reactions	$2\,j$
boundary conditions: $E(0) = 0,\ E(L+1) = 0$	2
steady state: $J_i(n) = J_i(n+1),\ n = 1\dots L-1$	$j \cdot (L-1)$
total	$jL + j + L + 3$

meters, p_{O_2}, T, L_{max}, the rate constants and equilibrium constants, as well as, the energetic parameters of the oxide phase considered are constants of the system and have to be defined in advance. The mathematical procedure provides a solution for all currents, the concentrations of all species and the electric field in the model system for the lattice planes 0 to $L - 1$ and at the surface L. The temporary oxide thickness is $l = a \cdot L$. The oxide growth curve $l(t)$ follows from the integration of the differential equation:

$$\frac{d[l(t)]}{dt} = \sum_{i=1}^{j} R_i J_i(l) \tag{5.31}$$

which means that the growth rate is proportional to the sum of all currents of atomic and ionic defects. Integrating (5.31) yields

$$t(l) = \int_0^l \frac{dl^*}{\sum_{i=1}^{j} R_i J_i(l^*)} \tag{5.32}$$

with the integration variable l^*. The oxide layer thickness $l(t)$ is obtained by inversion of (5.32). In principle, any system with numerous defects and currents can be solved. The proper choice of a model system is only a question of economy in mathematical efforts. Since each additional defect species introduces additional system constants, and consequently more free parameters, modeling of complex systems does not contribute much to a better understanding of the mechanisms. Introduction of more than the minimum number of defects may be helpful only if well documented experimental results cannot be interpreted by one of the simpler model types.

5.5 Example of a Model Considering Space Charges

The configuration of a simple model with a minimum number of defect species is shown in Table 5.3. Only one ionic defect and one chemisorption particle are assumed to be active and have to be considered and an oxide with the composition MO will be formed. The treatment of this first example is rather comprehensive since it includes also space charge effects in the oxide coating.

Table 5.3 List of defects and equations for a simple model including space charges

ionic defect	chemi-sorption particle	interface reactions, equation No	stoichio-metric factors
O^{z-}	O^-	MO formation: (5.6)	$x = y = z = 1$
		chemisorption: (5.11)	$\beta = \delta = 1$
		O^{z-} formation: (5.14)	$z = \beta = \delta = 1$

5.5.1 Equilibria of the Interface Reactions

Equation (5.6) with the parameters given in Table 5.3 yields an equilibrium defect concentration $\theta_{O^{z-}} = K_6^{-1} \exp[-ze\varphi(1)kT]$ or with the parameter $E(1)$ instead of $\varphi(1)$

$$\theta_{O^{z-}} = \frac{1}{K_6} \exp\left(-\frac{zeE(1)}{kT}\right). \tag{5.33}$$

The units are taken from Tables E.1, 2. The equilibrium expression for the reaction equation (5.14) in general form reads as

$$[\theta_{O^{z-}}(L)]^{\beta/z} [\theta_{O_\delta^z}]^{-1} [\theta_{O_2}]^{\delta/2} = K_{14}. \tag{5.34}$$

With the parameters given in Table 5.3 this can be written as

$$\theta_{O^{z-}}(L) = \theta_O - K_{14}. \tag{5.35}$$

5.5.2 Ion Current

The ion defect current between the planes n–1 and n is given by (D.1) or

$$J_{\mathrm{ion}} = D\left[\theta_{O^{z-}}(n)\exp\left(-\frac{E(n)}{2kT}\right) - \theta_{O^{z-}}(n-1)\exp\left(\frac{E(n)}{2kT}\right)\right] \tag{5.36}$$

for $n = 2$ to $n = L$. J_{ion} is assumed to be constant for all n values because of the steady state condition (5.19).

5.5.3 Electronic Currents

The relations used for electronic currents are defined in Appendix C. The partial forward and reverse currents for electron tunneling and for currents according to the hopping mechanism can be combined to a net electron current

$$J_{\mathrm{el}} = J_{\mathrm{tf}} - J_{\mathrm{tr}} + J_{\mathrm{thf}} - J_{\mathrm{thr}}. \tag{5.37}$$

This can be written as

$$J_{\mathrm{el}} = \theta_{O_2} F_f - \theta_{O^-} F_r. \tag{5.38}$$

F_f and F_r are functions of the potential difference V across the oxide layer and depend on the parameters W, U and m_e which characterize the electronic structure of the oxide system under consideration (Appendix C.1). The coverage of the surface with physisorbed oxygen, θ_{O_2}, is obtained from the condition that the physisorption forward flux (2.33) is balanced by the sum of the physisorption backward flux (2.34) and the production rate of chemisorption species. The latter term is in our present model proportional to $J_{\mathrm{el}}/2$. Inserting θ_{O_2} from this condition into (5.38) and solving the expression obtained for J_{el} leads to

$$J_{\mathrm{el}} = \frac{p_{O_2} K_p F_f - \theta_{O^-} F_r(1 + p_{O_2} K_p)}{1 + p_{O_2} K_p + F_f \beta/[(3-\delta)k_{1+}]}. \tag{5.39}$$

The potential V can be calculated from the values of the local electrical field $E(n)$. According to classical electrostatics one obtains

$$V = -\int_{n=1}^{L+1} E(n)\, dn \qquad (5.40)$$

If n is replaced by discrete values the integral (5.40) can be written as a sum

$$V = -a\sum_{n=1}^{L+1} E(n) \qquad (5.41)$$

and with the units given in Table E.1 ($a = 1$)

$$V = -\sum_{n=1}^{L+1} E(n). \qquad (5.42)$$

5.5.4 Mathematical Treatment

The aim of the mathematical procedure is the determination of the quantities electrical field, charge distribution in the oxide layer, the flux of oxygen interstitial ions as a function of the layer thickness given by the number of lattice planes,

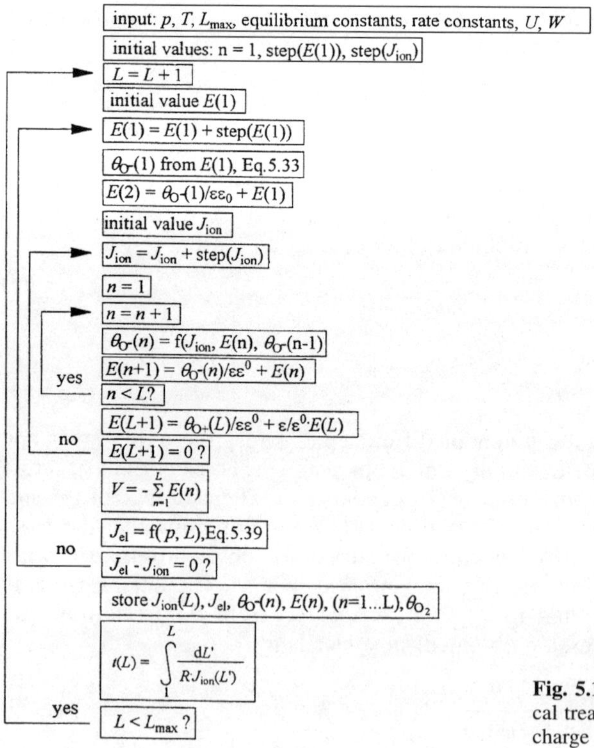

Fig. 5.10 Flow diagram for the numerical treatment of a model with the space charge inside the oxide layer

and finally by numerical integration, the oxide growth curve $l(t)$. The input parameters are the equilibrium constants K_6, K_p and K_{14}, the kinetic constant k_{1+}, the diffusion constant D for the defects, p_{O_2} and T. Two parameters, the initial field $E(1)$ at the plane $n = 1$ and the initial ion current J_{ion} are varied in two iteration loops until the charge neutrality condition is fulfilled which is checked by the condition $E(L + 1) = 0$ (Sect. 5.3.2) and until $J_{ion} = J_{el}$ holds according to the principle of coupled currents. The data obtained are stored in the computer together with the time needed by the ion flux to increase the oxide layer by one lattice distance a. The reaction time necessary to reach a thickness of L planes is obtained by integration of (5.31). This procedure is continued stepwise by increasing L until the maximum number L_{max} of lattice planes wanted in the model calculation is reached. The flow diagram of a computer program for the numeric procedure is shown in Fig. 5.10. The results of calculations for this model are presented together with results for a model neglecting space charges in Sect. 5.9.1. Comparison of both approaches shows that the chance to detect space charge effects merely by careful analysis of experimental oxidation curves is very limited.

5.6 Models Neglecting Space Charges

The space charges inside an oxide layer normally affect the reaction mechanism very little. They can be neglected if the sum of electric charges present in the bulk is small when compared with the charges on the oxide surface,

$$\sum_{i=1}^{j} \sum_{n=1}^{L-1} z_i \theta_i(n) \ll \sum_{i=1}^{j} z_i \theta_i(L) + \theta_{c(chem)}. \tag{5.43}$$

The condition $\theta_i(n) \approx 0$ inside the oxide phase has the consequence that according to (B.2); $E(n + 1) = E(n) = E_0$ holds for all planes n. Thus, no space charge means simultaneously homogenous electrical field inside the oxide layer.

5.6.1 Ion Current in the Homogenous Field

In the homogeneous field limit the ion current through the oxide scale is given by (D.16). In terms used in the present treatment it can be written as

$$J_{ion} = 2D \sin h\left(\frac{-z_i e V}{2kTL}\right)\left(\frac{\theta_i(L-1) - \theta_i(1)\exp\left(\frac{-z_i e V(L-1)}{kTL}\right)}{1 - \exp\left(\frac{-z_i e V(L-1)}{kTL}\right)}\right). \tag{5.44}$$

With $E_0 = V/(L)$ and units as given in Table E.1.

5.6.2 Electrostatic Phenomena

In a model without space charge only the charges present on the oxide surface and the compensating charges of opposite sign at the metal/oxide interface have to be considered. The system can then be represented by a plate condenser with the oxide layer as dielectric medium and with a surface charge

$$\sigma_{\text{surface}} = + \sum_{i=1}^{j} z_i \theta_i(L) - \sum_{\beta,\delta} \beta \theta_{O_\delta^{\beta-}}. \tag{5.45}$$

The potential difference across the oxide layer is given by

$$\frac{\epsilon_0 \epsilon_{\text{ox}}}{L} V = + \sum_{i=1}^{j} z_i \theta_i(L) - \sum_{\beta,\delta} \beta \theta_{O_\delta^{\beta-}}. \tag{5.46}$$

5.6.3 Surface Penetration

The models proposed assume that defects present inside the oxide phase are created and annihilated at the oxide surface $n = L$ and not at the subsurface $n = L - 1$. The surface penetration step is comparable to a diffusion jump. The main difference is a different activation energy. However, for charged defects the probability for the forward and reverse jumps are affected by the electric field. The net flux through the oxide interface is then given by

$$\begin{aligned} J_{\text{defet}} = v_+ - v_- = {} & \theta_{\text{def}}(L - 1)k_+^0 \exp\left(\frac{-A}{kT}\right) \exp\left(-\frac{z_i e V}{2kTL}\right) \\ & - \theta_{\text{def}}(L)k_-^0 \exp\left(\frac{-A}{kT}\right) \exp\left(+\frac{z_i e V}{2kTL}\right). \end{aligned} \tag{5.47}$$

5.6.4 Configuration of the Models

Models on oxidation kinetics have to be distinguished mainly with respect to the mobile ionic defects, namely, oxygen interstitials, O^{z-}, metal interstitials, M^{z+}, metal vacancies, $[M]^{z-}$, and oxygen vacancies $[O]^{z+}$. In Sects. 5.8–5.11 examples for each of the four point defects are presented. The main features of the oxidation curves calculated for individual defect types are similar but it will be demonstrated that the effect of parameter variations can differ considerably. For the mathematical treatment one has to define the sign of the currents. They are taken positive if they flow from the metal to the oxide surface. Thus, in oxidation systems the term $z_i J_{\text{ion}}$ is always positive. It compensates the current of electrons which always flow from the metal to the oxide surface because of the high electronegativity of oxygen atoms. This means that with $z_{\text{el}} = -1$ $z_{\text{el}} J_{\text{el}}$ becomes a negative quantity in the equation of coupled currents,

$$z_{\text{ion}} J_{\text{ion}} - J_{\text{el}}(V) = 0. \tag{5.48}$$

The potential V is negative for a negatively charged oxide surface. Data on the configuration of the four different models presented are listed in Table 5.4.

The principle of coupled currents (5.48) provides a very useful equation for the numerical treatment of the model calculations. This simple relation enables the

Table 5.4 Particles and equations used in models neglecting space charges

ionic defect	chemi-sorption particle	reaction Eq. o: oxide d: defect	factors	surface charge $\sigma_{surface} = \frac{\epsilon_0 \epsilon_{ox}}{L} V$
M^{z+}	O_2^-	(5.12) (o) (5.7) (d)	$\delta = 2$ $\beta = 1$	$-\theta_{O_2^-} + z\theta_{M^{z+}}(L)$
$[M]^{z-}$	O_2^-	(5.13) (o) (5.8) (d)	$\delta = 2$ $\beta = 1$	$-\theta_{O_2^-} - z\theta_{[M]^{z-}}(L)$
O^{z-}	O^-	(5.6) (o) (5.14) (d)	$\delta = 1$ $\beta = 1$	$-\theta_{O}^- - z\theta_{O^{z-}}(L)$
$[O]^{z+}$	O^-	(5.5) (o) (5.15) (d)	$\delta = 1$ $\beta = 1$	$-\theta_{O^-} + \theta_{[O]^{z+}}(L)$

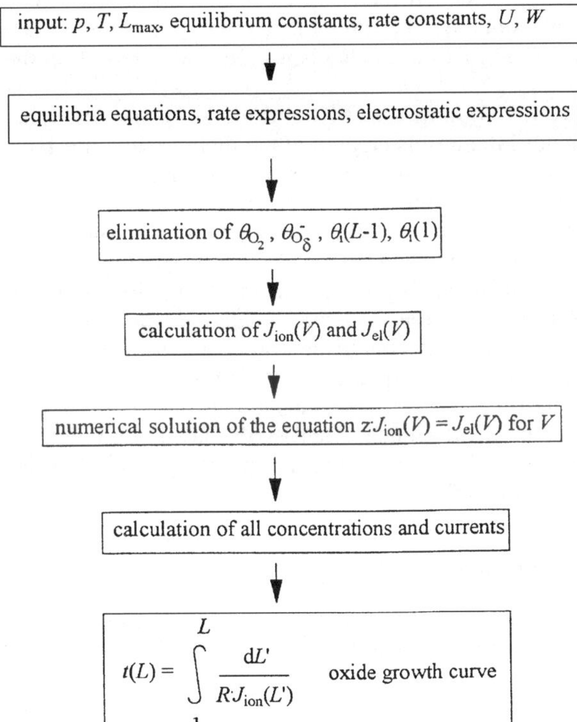

Fig. 5.11 Flow diagram for the numerical treatment of models without space charges inside the oxide layer

replacement of the electronic current by the ionic current and vice versa in all equations whenever this is convenient. Each of the variables occurring in (5.48) can be replaced by the potential difference V across the oxide layer if the set of equations available for reaction rates and equilibria is solved for each model by standard methods. Finally, an equation is obtained of the type

$$z_{\mathrm{ion}} J_{\mathrm{ion}}(V) = J_{\mathrm{el}}(V) \tag{5.49}$$

which can be solved numerically by the procedure shown in Fig. 5.11. After determination of the variable V all other data of the system can be obtained from the equations given for the actual layer thickness, L. The time integral over the flux of the ionic defects yields, again, the oxidation curve $l(t)$. With the set of data available for $L = 2$ to $L = L_{max}$ curves for all parameters of interest can be plotted as a function of time or layer thickness.

5.7 Detailed Presentation of a Model with Metal Interstitials as Mobile Defects

The mathematical procedure for the analysis of a model with metal interstitials M^{z+} and without space charge in the oxide layer follows the scheme presented in Fig. 5.11. The methods applied and the presentation of results are discussed in more detail only for this example since for the three other model types mentioned above they are about the same. The reaction mechanism proposed is based on the following assumptions:

The oxide formed shall have the composition MO and the ionic species transported through the layer are metal interstitials created at the metal/oxide interface. They migrate to the oxide surface and produce there, together with chemisorbed and physisorbed oxygen species, an oxide molecule. Thus, the reaction front is the oxide surface. Electrons are transported through the layer by tunneling and by the hopping mechanisms. Acceptor sites for electrons on the oxide surface are provided by physisorbed oxygen molecules. Physisorption and formation of metal interstitials are assumed to be fast partial reactions and treated as equilibria. The steady-state condition and the principle of charge neutrality are used whenever this is convenient. Therefore, the flux of any species occurring in rate or transport equations can be replaced by the flux of electrons times a constant factor, a term which depends only on the potential difference V across the layer. Last but not least, the values of the parameters defining the physical and chemical properties of the system must be chosen (Table 5.5).

5.7.1 Equilibrium and Rate Equations

The interface reactions used are listed in Table 5.5. The equilibrium condition for the interstitial formation obtained from (5.7) can be written as

$$\theta_{M^{z+}}(1) = K_7 \exp\left(\frac{-zV}{kTL}\right) \tag{5.50}$$

with $\varphi(1)$ substituted by V/L. $\theta_{M^{z+}}$ is given as a function of the potential V. The oxide formation reaction (5.12) yields the equilibrium equation

$$\theta_{M^{z+}}(L) = \left(\theta_{O_2}\right)^{(z-1/2)} \left(\theta_{O_2^-}\right)^{-z} k_{12}^{-1} . \tag{5.51}$$

In addition, the flux of metal interstitials through the oxide surface is needed. The term $J_{M^{z+}}$ in (5.47) can be replaced by zJ_{el}. The concentration of interstitials in the subsurface plane $\theta_{M^{z+}}(L - 1)$ is then a function of $\theta_{M^{z+}}(L)$ and V.

Table 5.5 Parameters of the standard curve for the M^{z+} model

parameter	symbol	value	
eqilibrium constants	K_7	10^{-7}	ML
	K_9	$7{\cdot}10^1$	ML/Pa
rate constants	k^0_{47+}	10^{13}	s^{-1}
	k^0_{47-}	10^{13}	s^{-1}
diffusion constant	D	$5{\cdot}10^5$	s^{-1}
band structure	U	1.1	eV
	W	0.9	eV
dielectric constant of the oxide	ϵ_{ox}	13	
conversion factor ion flux to layer growth	R	0.13	nm/ML
charge number of the interstitial ion	z	1	
electron mass ratio	γ_m	0.5	
activation energies:			
diffusion	A	6	eV
physisorption	A_{phys+}	0.25	eV
surface/subsurface	$A_{ch,ss}$	0.5	eV
subsurface/surface	$A_{ss,ch}$	0.5	eV
M^{z+} Formation	ΔG^0	0.4	eV
Temperature	T	333	K
oxygen pressure	p_{O_2}	1	Pa

5.7.2 Surface Charges

The equilibrium condition (5.51) and the correlation between surface charge and potential V (Table 5.4) are used to express $\theta_{O_2^-}$ and $\theta_{M^{z+}}$ as a function of the variable V only. Equation (5.51) inserted into the surface charge term yields

$$\theta_{O_2^-} - z\left[K_{12}^{-1}\left(\theta_{O_2^-}\right)^{z-1/2}\left(\theta_{O_2^-}\right)^{-z}\right] = \frac{\epsilon_o\epsilon_{ox}}{L}\,V. \tag{5.52}$$

This equation for the determination of $\theta_{O_2^-}$ is for $z = 1$ quadratic and for larger z values of higher order. However, since it is very unlikely that larger concentrations of positive metal ions and negative oxygen chemisorption species are present simultaneously on the oxide surface without forming oxide molecules the equilibrium constant K_{12} must be very large. Under this condition one can use the approximations

$$\theta_{O_2^-} = \left|\frac{\epsilon_0\epsilon_{ox}V}{2L}\right| - \frac{\epsilon_0\epsilon_{ox}V}{2L} \tag{5.53}$$

and by a similar treatment for $\theta_{M^{z+}}$

$$\theta_{M^{z+}} = \left|\frac{\epsilon_0\epsilon_{ox}V}{2L}\right| + \frac{\epsilon_0\epsilon_{ox}V}{2L}. \tag{5.54}$$

This means that only one kind of charged species is expected to be present on the oxide surface and the other one can be neglected.

5.7.3 The Potential V Across the Layer

The potential V is calculated using the principle of coupled currents (5.22). The ion current equation is given by (5.44) and the electronic current can be calculated from (5.39) by inserting $\theta_{O_2^-}$ instead of θ_{O^-}. The expressions for the concentrations $\theta_{M^{z+}}(1)$ and $\theta_{M^{z+}}(L)$ are functions of V. The final equation contains only V as variable, unfortunately however in linear terms as well as in exponential terms. Thus, the equation has to be solved by numerical methods.

5.7.4 Calculation of Concentrations, Currents and the Oxide Growth Curve

The numerical solution of the coupled current condition (5.22) provides a value for the potential V for given experimental parameters p_{O_2} and L. The values for all currents and concentrations can be calculated by inserting V in the same equations used above for the elimination of the concentration terms of intermediate reaction species. They determine the concentrations and fluxes for a given oxide-layer thickness L between the minimum value $L = 2$ and the maximum value chosen, L_{max}. The data obtained are stored as a file in the computer. The value $J_{ion}(L)$ for $L = 2$ to L_{max} is of specific interest. With these data for the ion flux the oxide growth curve is calculated using (5.32). The discrete values of $J_{ion}(L)$ obtained for steady-state conditions are used as data points in Newton's method of numerical integration of (5.32). For each value of the parameter L the oxidation time $t(L)$ is calculated by addition of the time intervals needed for the growth of each of the oxide layers $L^* \leq L$.

$$t(L) = \sum_{L^*=2}^{L} \frac{1}{2R} \left(\frac{1}{J_{ion}(L^*)} + \frac{1}{J_{ion}(L^*+1)} \right). \tag{5.55}$$

The data pairs $t(L)$ and L describe the oxide-growth curve $l(t)$ in units ML/s. To terminate a simulation curve at the oxide thickness L_{max} one has to stop the computer program after run $(L_{max} + 1)$.

5.7.5 Standard Oxide Growth Curve

A calculated oxide-growth curve is exhibited in Fig. 5.12. The absorbed amount of oxygen which is equivalent to the layer thickness given in ML O_2 is plotted versus the logarithm of the time. The potential V is shown in the lower part of the figure with the same time scale. The system parameters chosen for this curve are listed in Table 5.5. They have reasonable values which are compatible with their definitions and with typical experimental data and in accordance with estimates discussed in Chap. 2 and in the Appendix. This curve will be used as a standard curve for the demonstration of effects caused by parameter variations. The oxide-growth curve and the potential curve have a characteristic shape which can be subdivided into three stages. In *stage one* the growth is very fast and the potential is high. *Stage two* is characterized by a declining potential curve and a much slower oxide

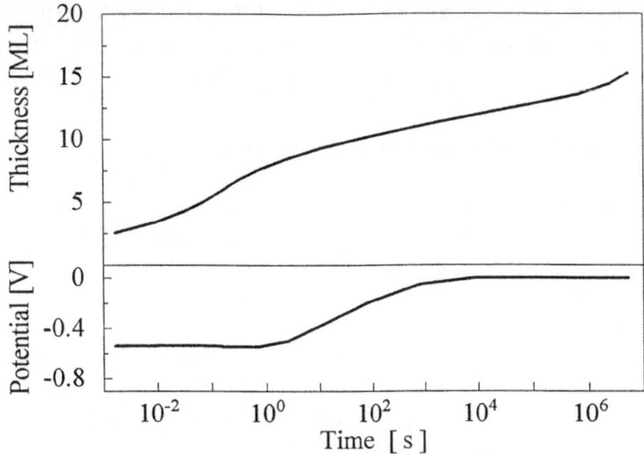

Fig. 5.12 Oxide growth curve and the potential V for the M^{z+} model with standard parameters

growth rate. In *stage three* the potential remains almost constant at a very low level. Discussions on details of the reaction mechanism in subsequent sections are closely related to the classification of oxidation curves into these three different stages.

The most important parameter for the interpretation of growth curves is the ion flux. Therefore, it has to be considered more closely. The equation used for the ion current in Table 5.5 can be simplified if the low field limit (D.21) is applied where the first term of a series expansion is used for $\exp(x) \approx 1 + x$ and $\sin h(x) \approx x$. This gives

$$J_{M^{z+}} = D\left(\frac{\theta_{M^{z+}}(1) - \theta_{M^{z+}}(L-1)}{L-1} + \frac{\theta_{M^{z+}}(1)zV}{kTL}\right). \tag{5.56}$$

The first term in large parentheses is the concentration gradient and the second the field term of the transport equation. The total flux is composed of a diffusion partial flux and an ohmic flux caused by the electric field V/L. $D\theta_{M^{z+}}(1)/(L-1)$ is a very important factor of (5.56) if $V \approx 0$ holds and $\theta_{M^{z+}}(L-1)$ is small compared to $\theta_{M^{z+}}(1)$. The concentration of M^{z+} defects at the metal oxide interface $n = 1$ is an almost constant quantity and given by the equilibrium condition (5.50). Thus, the factor DK_7 is a parameter which strongly affects the results. For the pre-exponential factor μ of D in (5.44) a value $v = kT/h \approx 10^{13}\mathrm{s}^{-1}$ is used according to Eyring's theory of the transition state (Sect. 2.2, [5.21]). A value of 10^{-7} ML has been chosen for K_7 that is comparable to the donator concentration in a Schottky barrier [5.22]. For the determination of the factor DK_7 reaction rates of experimental curves have been taken at the end of stage two, where the potential V is small. The ion current is given by the reaction rate times the factor $1/R$. This semi empirical procedure yields reasonable values for ionic currents in a model. Based on these estimates, the following parameters given in Table 5.5 have been chosen for the standard curve. The rate constants k_{47+}^0 and k_{47-}^0 of the surface penetration reaction (5.47) of M^{z+} defects and their activation energies must be compatible with the value of the diffusion constant. The pre-exponential factors for the transitions between the oxide surface and the subsurface have the same value as for bulk diffusion and in the standard curve also the activation energies are the same. That means the surface penetration is neither enhanced and nor impeded.

5.7.6 Concentration of Reacting Species and Partial Fluxes

The ionic and electronic partial fluxes, the concentrations of the reacting species at the interfaces and the potential V are plotted in Fig. 5.13 as a function of the oxide-layer thickness in units ML which are equivalent to lattice distances a of about $3 \cdot 10^{-10}$ m. The following processes are responsible for the reaction mechanism in the three stages of the oxidation curve:

The first stage is characterized by a high negative potential V, called here Mott equilibrium potential. This is a result of the equilibrium between free electrons in the metal and the electrons trapped by O_2^- chemisorption particles at the oxide surface. It can be sustained only at a relatively small thickness of the oxide layer as long as electrons can easily tunnel through the oxide layer, in our model up to 13 ML O_2. The ion current which is the rate determining step in this stage one of the oxidation process is equal to the small difference between the forward and the reverse tunnel flux. It is enhanced drastically by the equilibrium Mott potential V and the oxide-layer growth proceeds rapidly. With increasing thickness the potential V breaks down since electron tunneling is decreased strongly. Consequently, in stage 2 the ion current decreases strongly, too, due to declining field assistance of the ion flux. Now, the rate determining step is the forward tunneling flux which is balanced by the ion flux. In stage 3, the thermal electronic current plays the predominant role for charge compensation of the ion current since the tunnel currents approach negligible values. The weak field present guarantees charge neutrality according to the mechanism of ambipolar diffusion in semiconductors (Appendix D.3). The reaction rate is now determined by the slower of both partial fluxes, ionic or electronic.

The concentration of O_2 physisorption molecules on the surface of the oxide is in equilibrium with the gas phase and, therefore, independent of the layer thickness. The equilibrium Mott potential across the oxide phase stems from the ad-

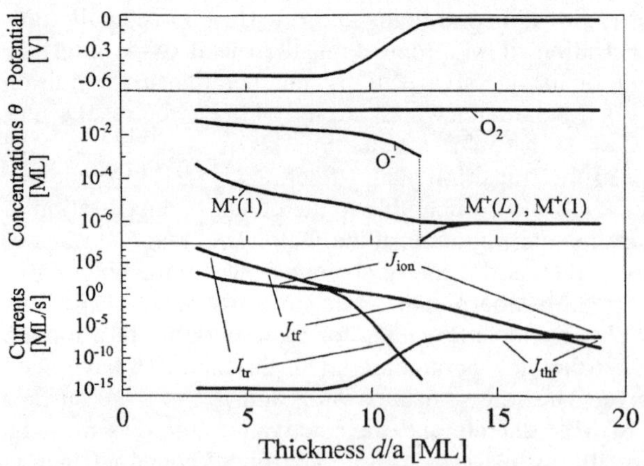

Fig. 5.13 Currents, concentrations and the potential V for the M^{z+} model with standard parameters.

sorption of O_2^- species. Their concentration decreases about proportional to $1/l$ as long as the potential V remains constant. After the potential is broken down the O_2^- particles disappear from the surface and few M^{z+} defects are present now to produce the very weak positive potential necessary for ambipolar defect diffusion in the oxide. The equilibrium concentration of M^{z+} defects in the oxide at the metal/oxide interface is field dependent and decreases with the decreasing potential.

The programs available run on PC computers and for each model with fixed parameters a set of plots as shown in Figs. 5.12, 13 is obtained within few minutes. Thus, this kind of computer simulations provides a convenient tool for extended studies on the behavior of individual model types. The analysis of parameter variations frequently indicates whether a specific mechanism may be suited to describe experimental curves on low temperature oxidation with realistic system parameters.

5.8 Results of Model Calculations, Parameter Variations

The standard oxidation curve for the model with M^{z+} interstitials in the oxide phase has the same shape as the experimental curves shown in Figs. 5.1–5.4. The extended branch of stage 1 appearing in the model curves for exposure times below one second cannot be measured and it is represented, therefore, in the experimental curves by a step. But in both cases the logarithmic stage 2 is found and is followed sometimes after several hours or at elevated temperatures by a stage 3 with a time law for the layer growth faster than logarithmic. The large number of system parameters available in the models (Table 5.5) makes it easy to fit any curve measured. Fitting of a single curve does not prove, of course, whether a mechanism proposed is realistic. More predicative are studies on effects of parameter variations in different model types. First in this section the M^{z+} model will be analyzed in some detail. Based on these findings deviations in the behavior of the other model types will be discussed.

If the influence of system parameters on the shape of simulated oxidation curves for the M^{z+} model are studied, a considerable number of quantities have to be checked, namely, the charge number of the ionic defect, the effective mass of the electrons in the oxide, some energy levels of the electrons in the system, and the experimental parameters pressure and temperature. The examples presented show that individual parameters frequently affect remarkably only one of the three branches of experimental curves. Thus, reliably measured experimental oxidation curves can provide data or estimates on the values of system parameters and indicate whether or not they are compatible with the physical and chemical structure of the system investigated.

5.8.1 Effective Charge of Metal Interstitials

The shape of growth curves and of the potential V for the z values 1, 2, and 3 are depicted in Fig. 5.14. Stage 1 becomes much shorter for $z = 2$ and 3 than for $z = 1$ and the beginning of stage 2 is shifted remarkably toward lower exposure times. However, the oxide layer thickness is almost the same after a very short exposure

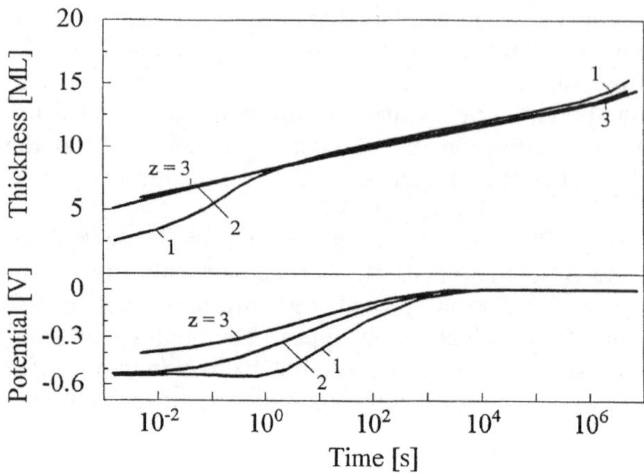

Fig. 5.14 Oxidation curves and potential V for different effective charge numbers z for the M^{z+} model

time for all three z values. Stage 3 begins after longer exposure time sooner for larger z values. The extension of stage 1 for $z = 1$ to higher exposure time is accompanied by a high Mott equilibrium potential V. For $z = 3$ the potential V attains not the plateau of the Mott equilibrium potential because of the very large value of the ion flux which can compensate the electron tunnel flux at electric fields below the Mott field (5.44). Equation (5.50) shows that not only the mobility of the interstitials is enhanced by z but also the number of defects by the exponential factor $\exp(+|zV|/kTL)$. This explains the strong dependence of the initial oxidation rate on the parameter z. However, since very fast rates in the initial stage 1 cannot be observed only the effects affecting stage 3 are of practical relevance.

5.8.2 Ion Current

Figure 5.15 shows oxidation curves and potentials for three selected values of the activation energy of diffusion. The same changes are observed for the product of the diffusion constant and the equilibrium constant for the formation of M^{z+} interstitials (5.7) as parameter. This term determines the size of the ion current if no electric field is present. Stage 1 is again strongly affected by the factor DK_7 since there the ion current is the rate determining step. For very high DK_7 values the ion current can become so high that the tunnel forward current can be compensated at a potential V below the Mott equilibrium potential and it is then rate determining. At very low ion fluxes the Mott equilibrium potential can be sustained to the point where the thermal electronic flux becomes equal to the tunneling current (Fig. 5.13).

5.8.3 Surface Penetration

If the forward flux of surface penetration is small when compared with the diffusion flux in the oxide then this partial step can become rate determining. Figure 5.16 shows such a situation where the surface penetration has been impeded by

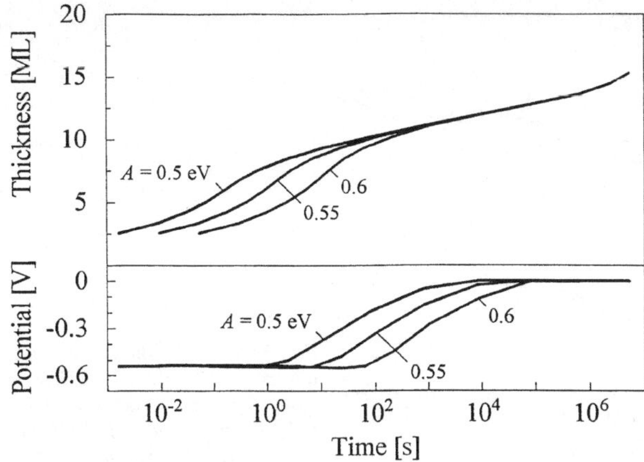

Fig. 5.15 Effects of the variation of the activation energy of the diffusion constant on oxidation curves and potentials for the M^{z+} model

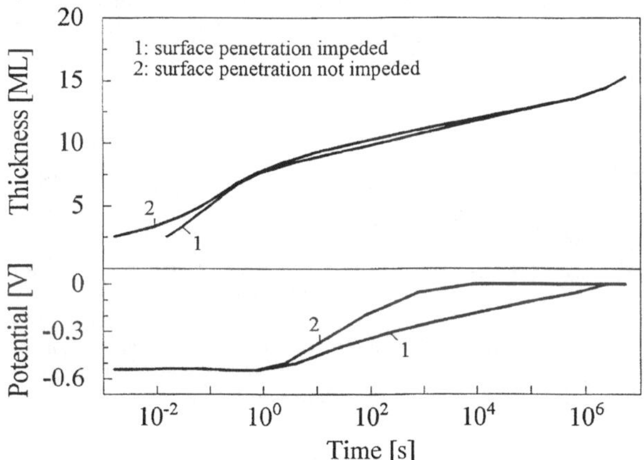

Fig. 5.16 Oxidation curve calculated with and without impeded surface penetration for the M^{z+} model

increasing the activation energy from o.5 to 0.6 eV. Larger differences are also observed here only in stage 1 where the ion flux is rate determining. Thus, the effect of this partial step is hard to detect by an experiment, too.

5.8.4 Effective Electron Mass

In Fig. 5.17 the ratio $\gamma_m = m_e/m_e^0$ has been varied. This parameter has a detectable influence on the oxide layer thickness in stage 2 because γ_m shows up as a factor in the exponential expression of the tunnel probability (C.2). Small values of γ_m increase the tunnel currents and extend stage 1, that means stage 2 begins with thicker oxide layers.

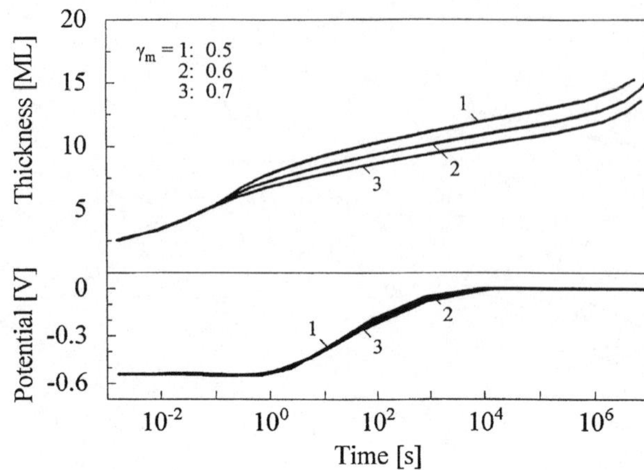

Fig. 5.17 Oxidation curves and potentials V for three different values of the ratio effective to the rest mass of the electron, $\gamma_m = m_e/m_e^0$ for the M^{z+} model

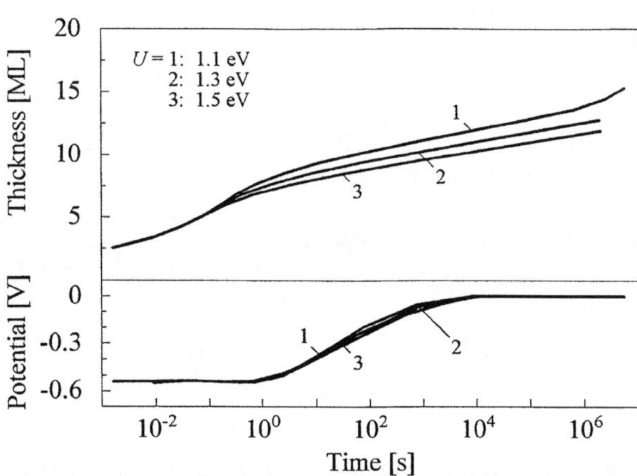

Fig. 5.18 Oxidation curves for different values of the energy gap U between the Fermi level in the metal and the bottom of the conduction band of the oxide for the M^{z+} model

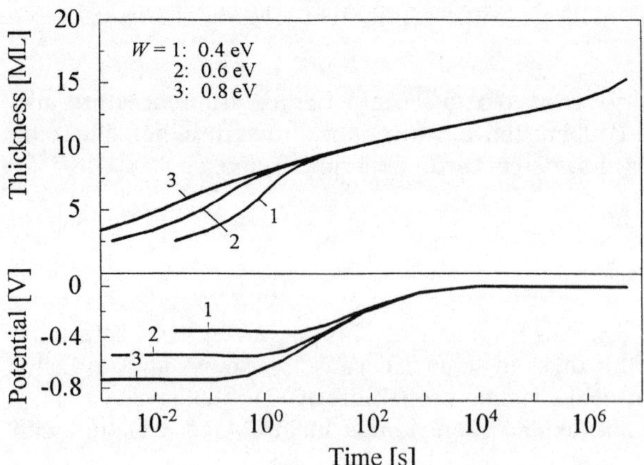

Fig. 5.19 Oxidation curves for different distance W between empty chemisorption levels and the Fermi level in the metal for the M^{z+} model

5.8.5 Energy U of the Conduction Band Distance

Effects of the variation of the energy gap U between the Fermi level in the metal and the bottom of the conduction band in the oxide are displayed in Fig. 5.18. Stage 2 of the curves is shifted to thicker oxide layers if the values of U are decreased. This is caused by a higher tunnel current. Smaller U values also increase the thermal electron current and shift the onset point of stage 3 to a smaller layer thickness. The latter effect is more pronounced than the first one.

5.8.6 Energy W of the Chemisorption Level

The effects of this parameter variation are displayed in Fig. 5.19. Higher W values increase the reaction rate in stage 1. The reason for this is the higher equilibrium Mott potential which enhances the ion current in stage 1 where it is rate determining.

5.8.7 Equilibrium Constant of Physisorption

The variation of the physisorption enthalpy ΔH_{phys} causes strong changes in the shape of the oxidation curves (5.20). ΔH_{phys} determines the concentration θ_{O_2} and the electronic forward fluxes (5.38). The higher the tunnel forward flux the higher is the Mott equilibrium potential and the longer it is sustained. Therefore, the branches 1 and 2 of the oxidation curves are shifted to higher thickness values. Since the thermal electronic flux is defined to increase proportionally to θ_{O_2} (C.9), the branch 3 also begins earlier. The statements made so far assume that the oxide surface is not saturated with O_2. If this is the case, the oxidation curve is no longer changed by K_{phys}.

5.8.8 Oxygen Pressure

Figure 5.21 depicts oxidation curves and the potential for four different values of the experimental parameter pressure, p_{O_2}. The effect of p_{O_2} on the oxidation behavior is the same as that of θ_{O_2} discussed above. If the limit at very low pressure is

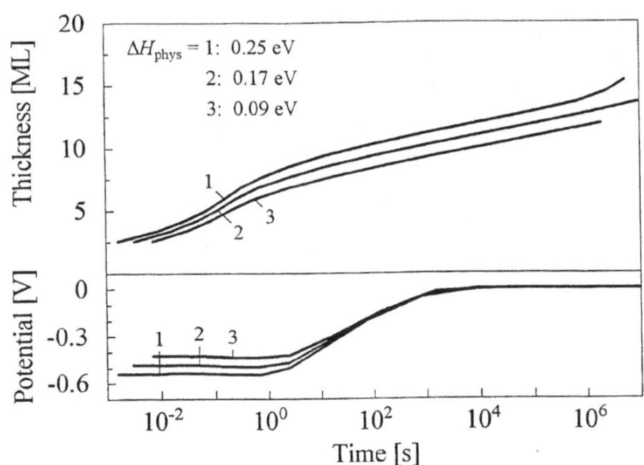

Fig. 5.20 Effects of the physisorption enthalpy ΔH_{phys} for the M^{z+} model

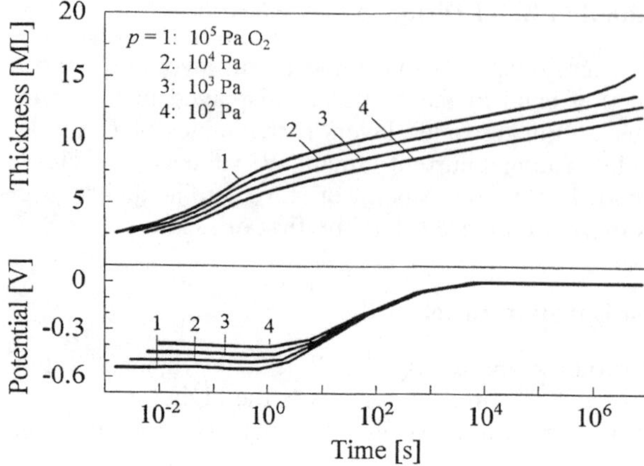

Fig. 5.21 Effect of the O_2 pressure on oxidation curves and potentials for the M^{z+} model

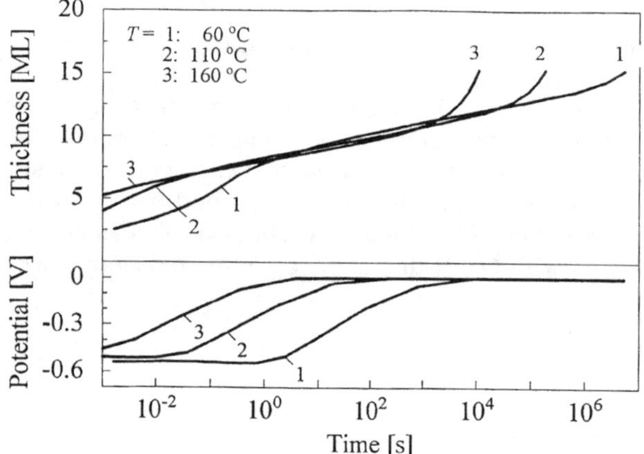

Fig. 5.22 Effect of temperature on oxidation curves and potentials for the M^{z+} model

disregarded, then θ_{O_2} is proportional to p_{O_2} and the electron currents as well as the ion current increase with increasing pressure. However, in systems where the condition $K_{phys} \cdot p_{O_2} \gg 1$ holds θ_{O_2} approaches the constant saturation value of 1 (A.20–22) and the oxidation curves become pressure independent. The saturation of physisorption sites or a fixed number of other acceptor sites for electrons at the oxide surface can be responsible for the small pressure dependence sometimes observed with experimental oxidation curves at high pressure, for example with nickel [5.13]. On the other hand, if systems with a fixed number of electron acceptor sites on the oxide surface are studied by a model this situation can be easily simulated by choosing K_{phys} big enough to get saturation. Then the models already available as computer programs do not have to be modified.

5.8.9 Temperature

Temperature enters as a factor in the exponential terms of almost all system parameters, rate and equilibrium constants of partial reactions as well as the mobility of the ion current. Consequently, a strong effect of temperature variation is observed in stage 1 of the curves shown in Fig. 5.22 because there the ion current is rate determining. Almost no difference in the curves is observed in stage 2 since the rate determining tunnel forward current is temperature independent. The thermally activated electron current is rate determining in stage 3 and there causes a strong temperature dependence of the oxide growth rate.

5.9 Effects of the Defect Structure of the Oxide

Some important details in the behavior of simulated oxidation curves depend on the type of the defect which is responsible as mobile ionic species for the transport of the reactants to the reaction front. If it is formed at the metal/oxide interface, such as M^{z+} metal interstitials or $[O]^{z+}$ oxygen vacancies, then cations migrate to the oxide surface. The formation is accomplished by the reaction of metal atoms with an oxide molecule. Thus, the equilibrium concentration of cations at the metal/oxide interface depends only on temperature and the electric field but not on the O_2 pressure. Since the defect flux is about proportional to the term $D\theta_{def}(1)/L$ it is mainly a function of T and V.

The anions $[M]^{z-}$ metal vacancies and O^{z-} oxygen interstitials are produced by the reaction of physisorbed and/or chemisorbed oxygen species on the oxide surface with oxide molecules in the bulk. Consequently, their concentration on the surface is a function of the O_2 pressure and of the temperature but not of the electric field. Therefore, the ionic flux $J \propto D\theta_{def}/L$ can depend on pressure according to the equilibrium constants given in Appendix A.2.

The argumentation made so far neglects the pressure dependence of the Mott equilibrium potential which can produce an indirect pressure dependence of θ_{def}. In contrast to the pressure dependence, the effects caused by variation of other parameters are less pronounced. The changes observed with oxidation curves calculated for the other three basic reaction models are about the same as shown in the examples of the preceding section for the M^{z+} model and must therefore not be discussed again.

5.9.1 Models with Oxygen Interstitials, Space Charge Effects

In Fig. 5.23 the standard oxidation curves for two models with O^{z-} interstitials as predominant ionic species are shown. The one is calculated under consideration of space charge effects in the oxide layer and the other without. This system is not very realistic from the point of view of material science because of the large size of oxygen ions, but for the mathematical treatment it has the advantage that the chemisorption and the interstitial species are identical. Therefore, this system has been chosen for the more lengthy numeric calculations of a space charge model. The parameters for both standard curves are very similar and in most cases identical with the data presented in Table 5.5. In Fig. 5.24 concentrations and fluxes for the constant-field approach are plotted versus the oxide-layer thickness. As with M^{z+} defects the mechanism can be subdivided into three stages. However, an important difference exists between the cationic M^{z+} model and the anionic O^{z-} model. In the O^{z-} model both defects present on the oxide surface have

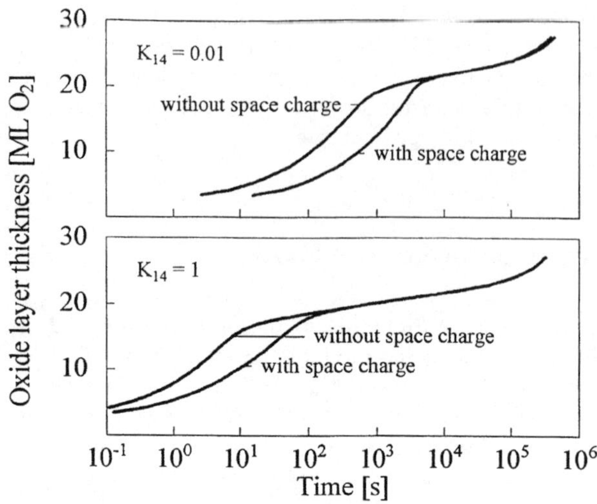

Fig. 5.23 Standard oxide growth curves for the O^- model with and without space charge effects. K_{14} is the equilibrium constant for O^- formation

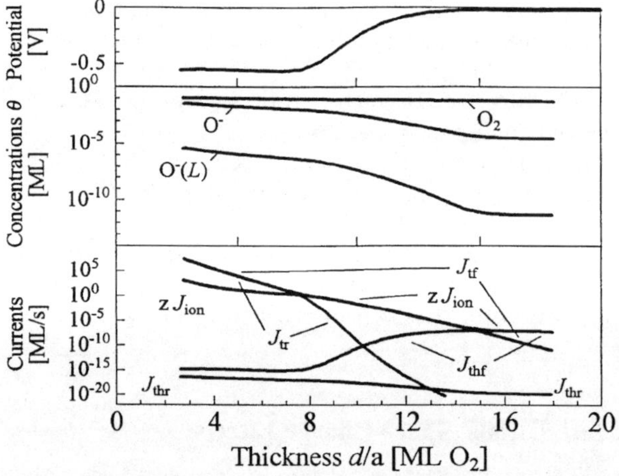

Fig. 5.24 Potential, concentrations and currents for the O^- model without space charge effects

a negative charge, the O^- chemisorption particles and the O^{z-} interstitials, whereas in the M^{z+} model positive interstitials or negative O_2^- species can be present. Thus, the potential cannot become positive. In anionic models the defects produced at the surface flow from the surface to the metal phase and, therefore, $\theta_{O^{z-}}(L)$ is much larger than $\theta_{O^{z-}}(1)$. According to (5.14) and the parameters chosen for this model the formation of O^- interstitials from O^- chemisorption particles does not depend on θ_{O_2}. Thus, a pressure dependence of the concentration of O^{z-} defects must be caused by the pressure dependence of the formation of O^- chemisorption particles. The concentrations of both species decreases strongly when the flux of electrons to the oxide surface is decreased during stages 2 and 3.

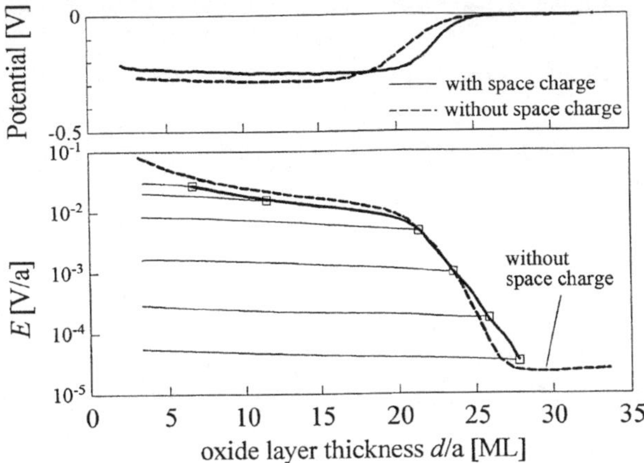

Fig. 5.25 Potentials and electric field for the O^- model with and without inclusion of space-charge effects

The effects of space charges inside the oxide phase on the shape of an oxidation curve is shown in Fig. 5.23. Consideration of space charges in the time consuming numerical treatment outlined in Sect. 5.5 causes only in a shift in the time scale of stage one by less than one order of magnitude to higher exposure time. This is caused by different potential profiles inside the oxide. Figure 5.25 presents a comparison of the field values obtained by both methods. The presence of the space charge reduces the Mott equilibrium potential. This has a similar effect as decreasing the value of the chemisorption level W (Fig. 5.19). Deviations of the homogenous field approach from the electric field of the space charge solution are small for the defect concentrations chosen in realistic quantities of about 0.1% or less. Nevertheless , the electric field and the concentration of interstitials, $\theta_{O^-}(L)$, in stage 1 and 2 are slightly smaller than without the space charge. This is an effect of the electric screening by the space charge. Comparison of both models demonstrates that small space charges which can be tolerated within the structure of the models affect the oxidation curve only in the initial phase of stage 1 where they cannot be determined experimentally. The differences occurring can be compensated easily by slight variation of the diffusion constant of the ionic flux. Since exact room-temperature diffusion data for point defects in oxides are normally not available, the constants inserted in models may be considered as reasonable estimates anyway. Simulations using the awkward numerical treatment necessary for the inclusion of space charge effects are mostly not very efficient. The much faster constant field approach is therefore highly recommended as a reliable approximation.

5.9.2 Effect of the Oxygen Pressure

Oxidation curves for three different oxygen pressures p_{O_2} are shown in Fig. 5.26 for the O^{z-} model with z = 1, as in the standard curves, and for z = 2 in Fig. 5.27. Increased pressures cause an increase of the Mott equilibrium potential due to an increased number of electron acceptor levels (5.38). Consequently, a parallel shift

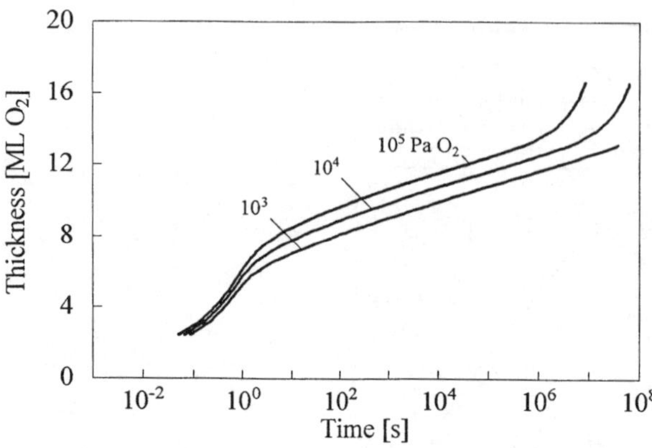

Fig. 5.26 Effects of the O_2 pressure on the oxide growth curves for a O^- model

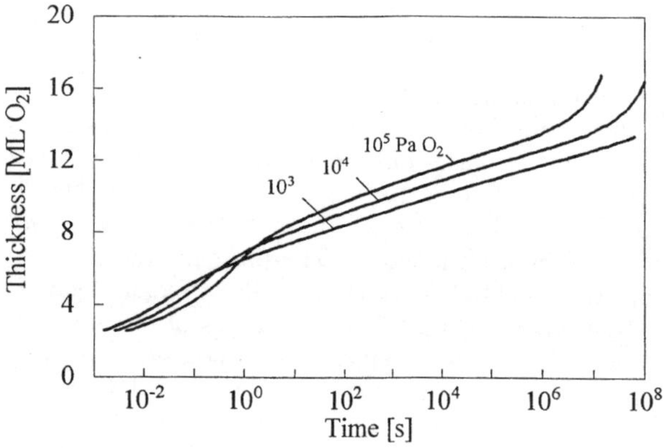

Fig. 5.27 Effects of the O_2 pressure on the oxide growth curves for a O^{2-} model

of the curves in stage 2 to higher oxide-layer thickness is observed. The deviation of stage 3 occurs sooner at higher pressures but at a similar layer thickness since the pressure dependence of the tunneling and the thermal partial flux is the same in our model. The changes in stage 1 depend on the defect formation reaction (5.14). If the reaction does not explicitly depend on pressure ($z = 1$), then the higher Mott equilibrium potential increases the reaction rate. For $z = 2$ the concentration of O^{2-} ions on the surface decreases with rising pressure. This effect overcompensates the effect of the Mott equilibrium potential.

5.9.3 Models with Oxygen Vacancies and Metal Vacancies

Models with $[O]^{z+}$ vacancies have a mathematical structure which is identical with the models for M^{z+} metal interstitials and models with $[M]^{z-}$ metal vacancies are equivalent to the models with O^{z-} oxygen interstitials. The results of these model types are, therefore, exactly the same as for their counterparts just presented and discussed in more detail elsewhere [5.11, 23]. Since only effects caused by cations or anions as predominant defects can be distinguished by the models and probably also by experimental oxidation curves, the presentation of further examples yields no additional information.

5.10 Simulation of Experiments with the Volumetric Method

In the preceding sections the oxidation kinetics under isothermal isobaric conditions has been discussed. It determines, for example, the amount of absorbed oxygen from the weight gain of a sample. In experiments with the volumetric Wagener method (Sect.3.2) oxygen flows from a reservoir through a capillary to a closed reaction volume and is absorbed there by the oxidation of a metal sample, for example, of a metal film. In the experiment the flux through the capillary is known and the absorbed gas amount is determined from the pressure change in the reactor volume. If this kind of experiments is simulated, one has to consider the oxygen molecules in the reactor as an additional state of the reaction mechanism. The adaptation of models already available to this slightly changed reaction mechanism is no big problem.

5.10.1 Reaction Model

The steady state conception is applied also for the pressure $p_{O_2(rec)}$ in the receiver

$$\frac{dp_{O_2(rec)}}{dt} = 0. \tag{5.57}$$

This means that the gas flux through the capillary is quantitatively absorbed by the sample, an approximation which is fulfilled at the very low pressures usually applied in the experiments. The flux through the capillary

$$J_{cap} = KL_{cap}(p_{O_2}^0 - p_{O_2(rec)}) \tag{5.58}$$

with L_{cap} conductivity of the capillary in ML/(s·Pa). The constant K is defined by the condition that the flux J_{cap} must be equal to the defect flux through the oxide scale necessary to increase the oxide layer by one lattice plane or one monolayer equivalent of the oxide M_xO_y,

$$KL_{cap}(p_{O_2}^0 - P_{O_2(rec)}) = \frac{y}{2x} J_{defect}. \tag{5.59}$$

Furthermore, the balance equation for the physisorption state has to be changed so that the steady-state condition holds:

$$[KL_{cap}(p_{O_2}^0 - p_{O_2(rec)}) + k_{1+}p_{O_2(rec)}](1 - \theta_{O_2(phys)}) - \theta_{O_2(phys)}k_{1-} = \frac{y}{2x} J_{defect}. \tag{5.60}$$

The concentration of physisorbed molecules is increased by two fluxes in a Wagener-type experiment: The direct gas flux from the orifice of the capillary and the impingement of gas molecules from the receiver atmosphere. All other transport and equilibrium equations remain the same as in the isobaric models and also the algorithm for the numeric solutions is not changed in principle. The only difference is the use of the pressure $p_{O_2}^0$ in front of the capillary instead of p_{O_2} as one of the experimental parameters of the system.

5.10.2 Results of Model Calculations

Figure 5.28 shows a simulation run for a volumetric experiment of a system with M^+ defects as mobile species inside the oxide. The reaction rate is plotted in terms of the reaction probability r versus the oxide layer thickness. The reaction probability is the absorption rate normalized by the impingement rate; $r = 1$ means all O_2 molecules hitting the surface are absorbed. The pressure rise in the receiver during the exposure run is shown, too, in Fig. 5.28. The parameters of the model are almost the same as given in Table 5.5. Changes have been made only for the physisorption enthalpy, 0.6 eV. The pressure $p_{O_2}^0$ was 10^{-2} Pa and the conductivity of the capillary $1.75 \cdot 10^{-2}$ ML/Pa s. In Fig. 5.29 the potential V, the concentrations of reacting defects and the partial fluxes are plotted in the same way as for the standard curves of isobaric simulation runs.

Up to eight Ml O_2 layer thickness, the limited O_2 flux has the consequence that the pressure in the reactor is almost zero and, thus, the reaction probability is unity. The physisorbed oxygen on the surface is quantitatively reacted and, therefore, the tunneling current has no acceptor sites and the Mott equilibrium potential cannot be established at first. Between 6 and 10 ML O_2 the tunnel forward flux and the ionic flux are balanced by the increasing potential. At a thickness equivalent to 10 ML O_2 the physisorption equilibrium is attained. Then, according to the exponential thickness dependence, the tunneling current decreases and with it also the Mott equilibrium potential. From that point on the behavior of the reaction mechanism is the same as for isobaric runs in stages 2 and 3.

This example demonstrates that it is no problem also to describe situations where the oxide growth is controlled by limited impingement rates with models of the same structure. It is not only possible to simulate Wagener-type volumetric experiments but also contamination reactions during high-vacuum treatments or thin-film deposition processes. The adaptation to a specific problem is accom-

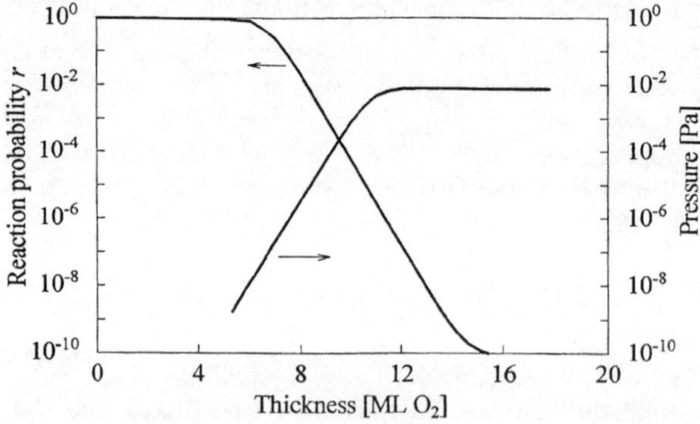

Fig. 5.28 Reaction probability, r, and pressure in the receiver for a simulation run of an experiment performed with the volumetric method (M^+ model)

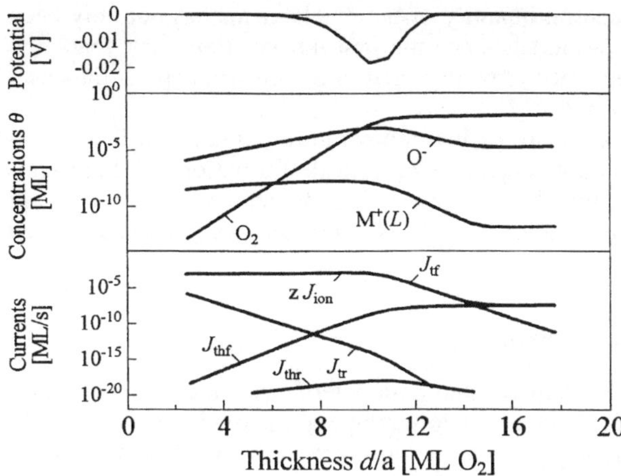

Fig. 5.29 Potential, concentrations of reacting species and fluxes determined for the model calculation presented in Fig. 5.28

plished by the definition of appropriate flux and balance equations for physisorbed molecules and the receiver volume which must be incorporated correctly into the main structure of the models.

5.11 Reaction Mechanisms of Low-Temperature Oxidation

In this section the conclusions derived from model calculations and experimental results will be emphasized. This may help proper rating of the slightly confusing variety of results and data presented in this chapter. Compared with the field of high-temperature oxidation, the understanding and theory of low-temperature oxidation had been fragmentary for various reasons. One of them are problems of experimental investigations mentioned in Chap. 3. They are finally responsible for the situation that many experimental results published in the past describe arbitrarily selected segments of the overall oxidation reaction. The data obtained had been frequently compared with semi empirical laws of models which assume that only one rate determining step is responsible for the oxidation kinetics observed. But the preceding sections of this chapter demonstrate that the mechanism of low temperature oxidation consists of three different stages. In each of them another partial step is rate determining. Unfortunately, the mechanisms cannot be idealized further than it has been done by the models presented without violating basic principles of physics and chemistry. The elements and basic processes of the individual partial steps involved are not new at all. New is only the consequent linking of a limited number of partial reactions in a model in order to obtain simulated oxidation curves for the whole process of low temperature oxidation. Comparison of experimental curves with computer simulations presented in this book and elsewhere [5.9, 11, 23] show that reasonable semi quantitative agreement can be attained without introducing undefined fitting parameters. The existence of the Mott equilibrium potential as the central quantity of the mechanisms has not only been proven by its successful use in models since the first introduction forty years ago [5.6] but also by measurements of the work function or of surface potentials as a function of oxygen exposure [5.24–27].

In the next section the mechanism of low temperature oxidation will be compared with the less complicated mechanisms of hydride formation and high-temperature oxidation. Then the functions of individual elements in the models are discussed and finally, it will be examined as to which conclusions can be obtained from experimental oxidation curves.

5.11.1 Fundamentals of the Mechanism

The relatively large number of elements and parameters needed, even for the presentation of the most simple models on low-temperature oxidation, can cause some perplexity. Therefore, looking at what is different from the models proposed for hydriding in Chap. 4 and the well-known oxidation models based on the theories of Wagner [5.3, 4] may be helpful. The elements necessary for development of simple, but self-consistent models are given in Table 5.6.

At first one has to distinguish systems forming metallic surface layers from those with an ionic structure. To the former belong most of the hydrides, carbides, and nitrides and to the latter oxides, sulfides or halogenides. In a metal matrix no electric field can exist and if ionic species would exist, screening by electrons from the conduction band would be completed within a very short distance so that all species adsorbed on the surface and defects inside the layer can be treated as neutrals. The growth mechanism can be simulated by the flux of a single uncharged defect through the layer. Depending on the parameters chosen a simulation run can supply results for a region where only one partial step is rate deter-

Table 5.6 Elements of modeling

element	hydride formation	high-temperature oxidation	low-temperature oxidation
fluxes	1, atomic	≥ 2, ionic, electronic	≥ 2, ionic, electronic
surface species	H_2(phys) , H(chem)	$O_2, O_\delta^{\beta-}$	$O_2, {}_\delta^{\beta-}$
defect in the layer	interstitials or vacancies	charged interstitials or vacancies	charged interstitials or vacancies
electric field	no	small ($< 10^7$ V/m)	high ($> 10^7$ V/m) *and* small ($< 10^7$ V/m)
potential ΔV across the layer	0	$< 0,05$ V	0.05–1 V *and* < 0.05 V
screening length λ	no	$\lambda \ll l$	$\lambda \gg l$
rate determining step	H_2 dissociation *and/or* diffusion	ambipolar diffusion	high-field migration *and* electron tunneling *and* ambipolar diffusion
change of steps	slow	no	rapid

mining or where two steps are relatively slow. Complete transition from one ideal region with only one rate determining step to another is normally slow and not observed in a single experiment.

If semiconducting layers are formed an electric field can exist inside the coating and the reacting species are normally charged. A fundamental requirement of all models with semiconducting layers is the condition of charge neutrality. In steady state the ionic flux responsible for the transfer of the reactants through the layer to the reaction front has to be compensated by an electronic current of equal size. This is accomplished by an electric field which retards the faster of both fluxes and enhances the slower one. If dry oxidation is compared with electrochemical processes one has to keep in mind that in the latter the electric field across the electrolyte phase is controlled by the experimenter as one of the experimental parameters whereas in dry oxidation it is an internal parameter which has to be adapted by the reaction mechanism.

The models on high-temperature oxidation consider the situation that the oxide layer is much thicker than the screening length of space charges which compensate interface charges on both sides of the oxide layer. In the bulk of the oxide no space charges exist and a very low electric field suffices to adapt the size of the ionic and electronic currents to the condition of charge neutrality. This phenomenon is called ambipolar diffusion. Since diffusion through thick oxide layers is normally much slower than interface reactions, the mechanisms of high-temperature oxidation is in most cases diffusion controlled.

Now, what is different with low-temperature oxidation? This is mainly the thickness of the layer which is much smaller than the screening length of charged particles in the oxide, at least in the initial stage. Under this condition the charges at the oxide surface and at the metal/oxide interface can react with each other and create a similar situation as in a plate condenser. The big difference of the electronegativity of oxygen atoms and metal atoms of about one eV is the reason for the existence of the Mott field which enhances the ionic fluxes by orders of magnitude in stage 1. The initially high tunneling current is strongly thickness dependent. Therefore, it becomes rate determining in stage 2 and has to be replaced in stage 3 by an ambipolar diffusion mechanism which is about the same as in high-temperature oxidation.

The comparison of the structure of the three reaction mechanisms shows the qualitative differences. Whereas the models for hydriding or high-temperature oxidation with a single rate determining partial step are able to describe real systems by consequently structured models, at least under specific ideal conditions, this is principally impossible for low-temperature oxidation. Here the overall process occurring on clean metal surfaces in oxidizing atmospheres can be simulated only by models which describe three stages, each with another rate determining step. The individual processes for the characterization of the phenomena mentioned in this section are explained in more detail in relevant sections of the Appendix.

5.11.2 Physisorbed Oxygen

Molecular oxygen must be dissociated before it forms an oxide and this process occurs at the oxide surface. The simplest way to accomplish this process is the introduction of molecular physisorption. The concentration of physisorbed oxygen molecules is proportional to the O_2 pressure. If the physisorption energy is high, the surface will be saturated at higher pressure or lower temperature. This has the consequence that a pressure dependence exists for reactions where physisorbed oxygen acts as a reactant and provides acceptor sites on the oxide surface for electronic currents. If the physisorption energy is small and the temperature relatively high the concentration is extremely small, also at high pressures. The physisorbed or molecular adsorbed states are the links between the gas phase and the processes on and in the oxide phase. They are filled from the gas phase and consumed by subsequent partial reactions. Correct incorporation of molecular adsorption states into a model can be checked by inserting very low pressure values. The reaction rate must then be determined by the impingement rate.

5.11.3 Defect Formation and Oxide Formation

For a larger number of oxides the defect structure is known or can be estimated. Since the negatively charged oxide surface acts as cathode for the ion transport only the concentration of anions, produced at the surface, can be pressure dependent whereas the concentration of cations, formed at the metal/oxide interface, is not directly pressure dependent. But the defect-formation reactions can depend also on the electric field at both interfaces and this depends on the paremeters $\theta_{O_2(phys)}$ or p_{O_2}. Therefore, the pressure dependence of an oxidation curve is no reliable indication whether the migrating defects are anions or cations.

5.11.4 Electronic Structure and Electronic Currents

The value of the Mott equilibrium potential depends on the energy level of the chemisorption particles, W. The thickness where it breaks down in the models depends on the value of the tunnel current. This current is increased exponentially with the barrier height $U^{1/2}$ and the effective electron mass. These two parameters determine the layer thickness after stage 1. Electron tunneling is not expected to yield currents high enough to sustain the Mott field for layers thicker than about one nm. Since some of the transition metal systems have an extension of stage 1 beyond this value, other electron conduction mechanisms have to be considered, too, for instance ohmic conduction through very thin films (Sect. C.3). If tunneling is defined to be the fast electron transport mechanisms, which is responsible for the development of the Mott equilibrium potential in the initial stage 1, then in addition a second conduction mechanism must be introduced which provides an electronic current through the layer at the final stage 3 according to the conception of ambipolar diffusion.

5.11.5 Effects not Considered by the Models

The structure of the models discussed so far are restricted to the absolute minimum number of elements and the elements itself are described by very simple equations. The most severe deviation from reality may be the assumption of an ideal and time-independent defect structure of the oxide layer. The oxides formed at low temperature usually have an amorphous structure [5.12]. They are in a metastable state and, thus, have the tendency to undergo ordering or relaxation processes. Therefore, the defect concentrations may be functions of time and of other parameters. The mechanisms of the electronic transport active in stage 1 may also be highly dependent on the structure of the oxide. Electrons can tunnel directly from the metal bulk to the oxide surface. But they can also tunnel from one acceptor site to another inside the layer. Another restriction is the limitation of models to one defect and one chemisorption species. If the pressure or temperature is changed or ordering processes proceed in the oxide not only the concentration of the migrating defect can be changed, but it can also happen that other defects take over the leadership in the transport mechanism.

5.11.6 Approximations for Estimated Oxidation Curves

For each of the three stages the rate determining flux can be approximated by relative simple equations. The overall rate curve is obtained by taking the lowest curve each. For stage 1 the ion flux can be written as

$$J_{\text{ion}} = D\theta_{\text{def.max}}(1 \; or \; \text{L}) \exp\left(\frac{zW}{2\text{L}kT}\right). \tag{5.61}$$

The Mott equilibrium potential has been replaced here by the value of the chemisorption potential W. For the presentation of stage 2 the tunnel forward current is approximated by [5.11]

$$J_{\text{tv}} = 2\theta_{O_2} \frac{U}{4\pi^2\hbar a^2\text{L}^2} \exp\left(\frac{-2\sqrt{2mU}a\text{L}}{\hbar}\right). \tag{5.62}$$

Stage 3 is represented by (C.9) with $V = 0$. Figure 5.30 demonstrates that the general feature of the original curve is approximated fairly well. This procedure can be used to find suitable parameters for fitting of experimental curves.

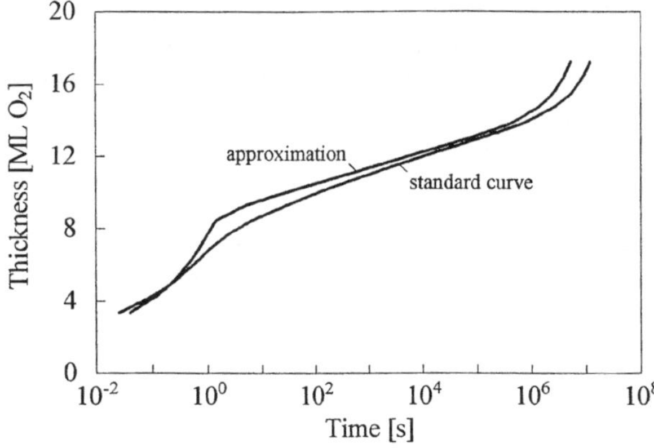

Fig. 5.30 Comparison of the approximation curve with model calculations for the O$^-$ model

5.11.7 Experimental Results and Model Calculations

The three-stage reaction mechanism of the models presented are well suited to describe and understand the general feature of low-temperature oxidation reactions. What can be checked by experiment are the thickness of the layer in the extremely fast stage 1 and the pressure and temperature dependence of the oxidation curves in the later stages.

Layer thickness
The layer thickness after stage 1 amounts from 1 to 2 ML for aluminum up to 10 to 100 ML for iron [5.14]. It depends on the capability of the electronic current to maintain electronic equilibrium across a growing layer. If only tunneling is considered metals with a large bandgap U between the Fermi level in the metal and the conductivity band in the oxide should form thin layers. The behavior of the aluminum oxidation seems to support this argument. Thick layers of some ten ML are found with film samples of transition metals. In the case of iron, the thickness measured with the quartz microbalance has been confirmed by transmission electron microscopy [5.14]. For tunneling this is an extremely large distance. In addition to the film roughness, also alternative fast transport mechanisms for electrons, mentioned in the Appendix C, can explain this result. Transition metals such as V, Ti, Fe, and Cr are especially capable to form suboxides with resistivity values which may be able to sustain a Mott equilibrium potential over larger distances by semiconduction [5.28, 29]. Extremely thick oxide layers are mainly found with transition metals.

Pressure dependence
The layer thickness attained after stage 1 is for some metals, for example vanadium, strongly pressure dependent and for others, such as nickel is not. At a first glance it appears like this has to do with the structure of the oxides formed. *Fehlner* [5.12] classified the metals according to the structure of the oxide skins formed at low temperature into network formers with anionic defect currents and network modifiers with cathodic fluxes. He and other researchers assumed that the pressure-dependent defect formation of anions at the oxide surface is the reason for this difference. However the discussion of pressure effects in our models shows that the situation is more complex. The pressure dependence can also be a problem of acceptor sites for electrons necessary to form charged chemisorption particles from physisorbed oxygen molecules. An explanation for strange-looking pressure effects could be also changes of the defect structure of a layer after the first seconds due to relaxation processes in the oxide layer. The two curves shown in Fig. 5.5 support this argument. The lower curve starts with an exposure pressure of 10^{-2} Pa. Then, the pressure was increased after each hour by one or two orders of magnitude. The increase of the layer growth is very small when compared with the oxide layer thickness of a film which had been exposed from the beginning to 10^2 Pa. In the models developed here the defect concentration depends only on the potential and the pressure but not on the structure which may change with exposure time. Therefore, the lower curve should have approached the upper one according to our rigid models. This demonstrates that more realistic models should consider relaxation effects, too. Figure 5.31 presents curves of a O^- model which fits experimental vanadium curves. It is no problem to get fair agreement with the

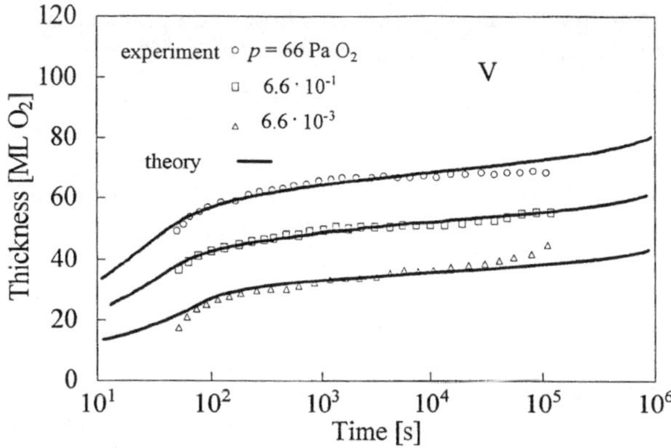

Fig. 5.31 Fitting of experimental vanadium oxidation curves for three different exposure pressures by an O$^-$ model

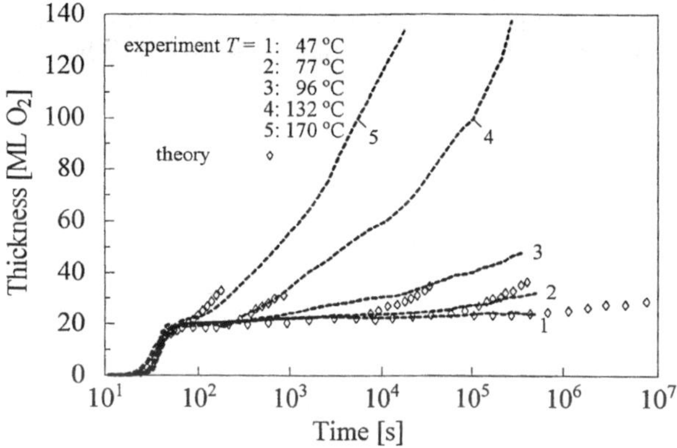

Fig. 5.32 Fitting of experimental iron oxidation curves for five different temperatures by a M$^+$ model

set of experimental curves due to the large number of parameters contained in the model. For a more detailed analysis of the reaction mechanism of such a system, further experimental data and the results of computer simulations with more advanced models should be available.

Temperature dependence

According to the models temperature variations affect only stage 1 and stage 3. Stage 2 is defined by a rate determining tunnel current which does not depend on temperature. The general shape of experimental curves agrees with the results of modeling. According to models the beginning of stage 3 has an onset point on the branch 2 of the curve which is shifted to lower exposure times with rising temperature. In contrast to this result of modeling, the experimental curves measured so far have their onset point always at the end of stage 1. This finding indicates again that the models are still too simple for quantitative simulation of all details of a reaction mechanism. Maybe the inclusion of semiconduction for the electronic flux in stage 1 would yield better fitting. Figure 5.32 shows computer simulations which represent the experimental curves measured for iron [5.14].

5.11.8 Conclusions

The mechanism of low-temperature oxidation of metals is a process with abrupt changes between at least three branches with different rate determining steps. Besides an enormous fund of qualitative information available on the thickness of oxidic tarnishing layers on metals and on the very initial stages of the process studied by surface analytical methods also in the last years quantitative data on reaction kinetics have been published. Such oxidation curves can be simulated by the type of models presented. In the present state of model development the elements incorporated have been restricted to processes which are represented by very simple equations. But even by these prototypes most of the general features of low-temperature metal oxidation can be studied and analyzed by appropriate parameter variations. It is no big problem to adapt the models to the simulation of a problem which may be of specific interest from a theoretical point of view. Thus, the present approach of modeling the reaction mechanisms in metal-gas systems provides a flexible tool for studies in the puzzling but fascinating field of low-temperature oxidation.

6 Poisoning of Hydrogen Reactions

Poisoning of metal-gas reactions at ambient temperature is a wide-spread phenomenon but the word poisoning is used for the characterization of quite different processes which retard the rate of a reaction. In catalytic processes, for example, molecules from the adjacent liquid or gas phase are adsorbed on a solid surface. There, with other species new molecules are formed as reaction products which are then released back to the liquid or gas phase. The reactivity of a surface acting as a catalyzer depends on active sites where the intermediate adsorption products are formed. These sites can be specific sites such as steps or adatoms or all sites on a clean surface. If they are blocked by immobile chemisorption compounds, the reaction is *poisoned*.

In this chapter only a very special aspect out of the large field of poisoning phenomena will be considered. This is poisoning of the hydrogen absorption reaction of metals by oxide layers or related processes. This is, roughly spoken, a combination of the effects discussed in the Chaps. 4, 5. If the metal surface is covered by a closed, dense oxide layer, as it is under normal conditions, the hydrogen molecules are dissociated at the oxide surface and hydrogen atoms or ions have to migrate as impurities through the oxide scale before they reach at the metal or the hydride phase. The reaction mechanism is very similar to the mechanism of hydride formation. The only difference in the case of hydride formation on a poisoned metal surface is the fact that not one but two surface layers have to be passed by the hydrogen atom before it arrives at the reaction front, which is the $/\beta$ interface (Fig. 4.3).

It has been demonstrated by the examples presented in Sect. 4.1 that hydrogen can penetrate clean metal surfaces very easily. However, if the thickness of the oxide layers exceeds a critical value the reaction rate is normally retarded by many orders of magnitude. A central problem of poisoning is the question: Which one of the two additional reaction partial processes is responsible for the impediment of the absorption reaction? Is it dissociation of H_2 molecules on the oxide surface or permeation of hydrogen atoms through the oxide layer? This question is also of high practical relevance if poisoning must be reduced or avoided, for example, in open hydrogen storage systems using metal hydrides as absorber materials. If poisoning is caused by slow H_2 dissociation one has to think about catalyzers on the oxide surface, and if it is permeation through the oxide layer one has to improve the hydrogen diffusion mechanism inside the oxide.

6.1 Experimental Results

Some general trends of poisoning effects during hydrogen absorption will be demonstrated by results obtained with wire samples, metal films and powders of hydrogen storage materials in various pressure regimes. The examples chosen are typical for the qualitative behavior of many other metal-gas systems investigated with the same techniques [6.1–13].

6.1.1 Wire Samples

Figure 6.1 shows the reaction probability of hydrogen for carefully outgassed tantalum wires, measured by the volumetric Wagener method under ultra high vacuum conditions as a function of the absorbed hydrogen content [6.1]. The curves for the clean surface and for surface contaminations below 7 ML O_2 begin with reaction rates close to the maximum r value of one where each H_2 molecule hitting the surface is reacting. After absorption of about 0.1 at-% H the reaction probability is reduced to $5 \cdot 10^{-2}$ which is still an extremely high value and it remains then almost constant. If the oxide layer thickness is less than about 7 ML O_2 poisoning of the hydrogen absorption process is not severe. However, for a layer thickness of about 14 ML O_2 the reaction rate is reduced drastically, the sample is *poisoned*.

Another example of poisoning effects is the hydrogen absorption of niobium wire samples measured by the change of the electric resistivity at 307 °C (Fig. 6.2). The experiments were performed under less clean surface conditions in a high vacuum system with a residual impurity-gas pressure less than 10^{-10} bar. Before each absorption run the sample has been outgassed at this pressure at 2 000 °C. The absorption curves in Fig. 6.2. a show that the first run and the fifth run after air exposure for three minutes at 40 °C are much slower than the runs two to four, where the receiver had not been floated with air between the indivi-

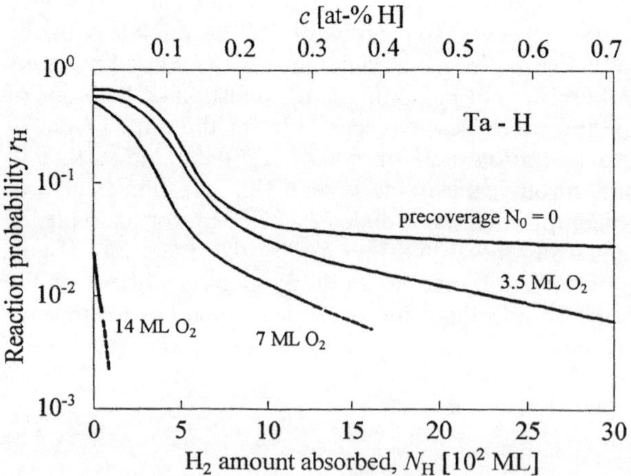

Fig. 6.1 Hydrogen absorption of a tantalum wire without and with oxygen precoverage in a plot reaction probability, r vs. amount of absorbed hydrogen N_H [6.1]

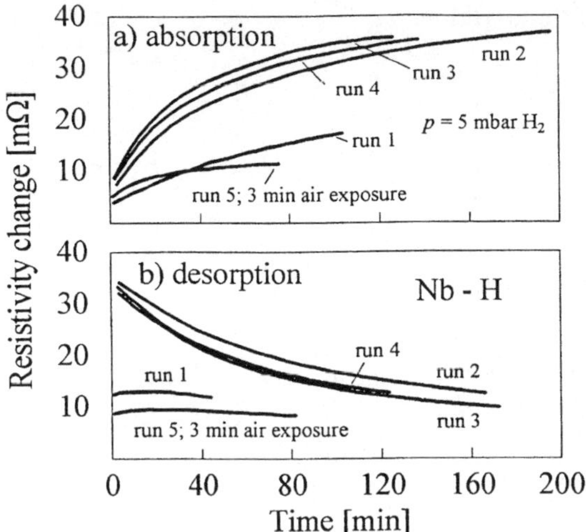

Fig. 6.2 Loading/unloading cycles for a niobium wire sample exposed at 600 K to 5 mbar H_2 . Before run 5 air exposure for 3 min at 320 K [6.14]

dual absorption and desorption experiments. The same behavior is also found with the dehydriding runs shown in Fig. 6.2. b. Also here the reaction rate for the runs one and five performed under slightly less clean surface conditions are much slower. However, it is very important to realize that even the faster reaction rates for the runs two to four are much slower than those found in the experiments with the clean metal surfaces shown in Fig. 6.1. There the reaction probability is $1 < r < 10^{-2}$ whereas the reaction rates presented in Fig. 6.2 correspond to r values in the range $10^{-6} > r > 10^{-8}$. This demonstrates that most results published on experiments performed under *high purity high vacuum conditions* are measured with samples having poisoned surfaces.

This situation is also demonstrateded by the data presented in Fig. 6.3. There the ranges of the initial reaction rate for tantalum wire samples are shown which have been coated with various metals acting as catalyzers. The full line on the top is the maximum rate expected for unpoisoned samples and a diffusion controlled absorption rate. The two lines at the lower left end of the figure represent curves for uncoated tantalum wires carefully and less carefully outgassed under high vacuum conditions. Palladium coatings on the tantalum wire prevent poisoning to large extent if the sample surface is not oxidized before coating or the palladium coating had been exposed to oxygen between coating and hydriding [6.3]. Iron, nickel, and copper show catalytic enhancement, too, but the effect is less pronounced. If all curves in Fig. 6.3 are extrapolated down to ambient temperatures near 300 K the difference in the reaction rates, or for the reaction probabilities, amounts to many orders of magnitude.

These examples demonstrate that the kinetics of hydrogen absorption and desorption of transition metals at ambient and slightly elevated temperatures is normally less a problem of bulk diffusion than a problem of defined surface states at the beginning and during the experiment. An increase of the outgassing temperature of hundred K, the change of the residual gas pressure by a factor of two or

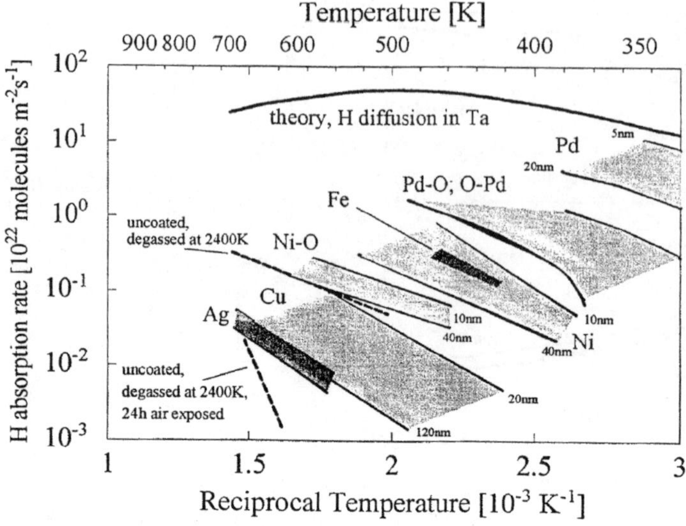

Fig. 6.3 Initial hydrogen absorption rate of coated and uncoated tantalum wires exposed to 50 mbar H_2 [6.15]

extension of the time between outgassing and exposure run by few minutes can change the thickness of the oxide surface layer drastically. These parameters, frequently not controlled quantitatively in an experiment, affect the kinetics at ambient temperature much more than the parameters pressure and temperature during hydrogen absorption which are normally given as characterization of experimental curves. The large scatter of data published by different authors on the kinetics of low temperature metal gas interactions is, therefore, a quite normal phenomenon.

6.1.2 Film Samples

Quantification of poisoning effects is a difficult experimental problem. The change of the hydrogen absorption rate has to be measured as a function of the oxide layer thickness. Since the thickness of the layers exceeds the monolayer range, in some systems even considerably, the sensitivity of most surface analytical techniques is not high enough for this kind of experiments. Poisoning effects can be measured by the modified volumetric Wagener method where partial pressure gages and multilayer deposition techniques are applied. This kind of experiments uses repeated deposition of one or several metal layers combined with hydrogen and/or oxygen exposures and provides quantitative, or at least semiquantitative, informations [6.4].

Figure 6.4 exhibits hydrogen absorption curves for titanium films with different precoverage of oxygen in a plot reaction probability, r_{H_2}, versus amount of absorbed hydrogen, N_H. The curve without precoverage looks very similar to the curve of the unpoisoned tantalum wire (Fig. 6.1). At first the maximum absorption rate at $r = 1$ is observed. After saturation of the α-metal phase β-hydride is formed on the film surface and the reaction probability is reduced to a constant value of about 10^{-2}. With increasing precoverage of the film surface, equivalent to

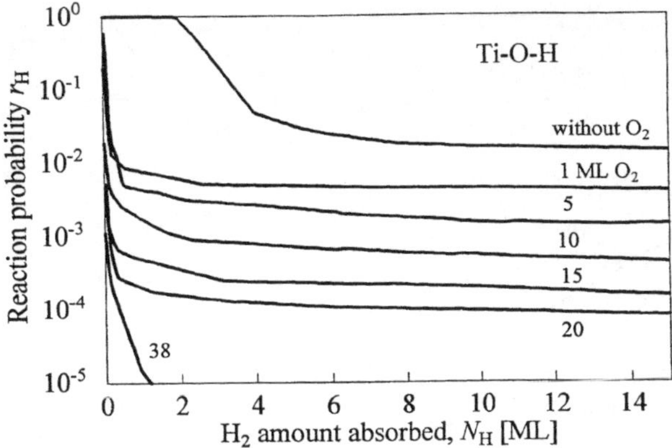

Fig. 6.4 Reaction probability for hydrogen on a 30 nm thick titanium film without and with oxygen precoverage [6.4]

1 to 20 ML O$_2$, the reaction rate is lowered by about two orders of magnitude, depending on the layer thickness. However, after increasing the layer thickness to 38 ML O$_2$ the reaction rate is drastically reduced and the sample can be characterized as being *poisoned*.

After the formation of surface hydride on clean titanium films the reaction probability of hydrogen does not depend on the thickness of the hydride layer, which is proportional to the amount of absorbed hydrogen, and the reaction rate is proportional to the exposure pressure [6.9]. This demonstrates that under the experimental conditions chosen the rate determining step of the process is the dissociation of adsorbed hydrogen molecules on the surface of the hydride or on the oxide layer and not permeation through the layers. Otherwise $r \sim 1/\sqrt{p}$ should have been observed.

Many investigations have been performed with this type of experiments. The results can be subdivided into four categories:

(i) In one group of experiments poisoning of different metals or storage materials by oxide layers has been measured [6.6–11]. The trends observed with films of Ti, La, Fe, Ni, Mn, Pd, Cr, V, FeTi, LaNi$_5$, TiMn$_2$, TiNi, and TiPd can be characterized by the following phenomena: Metals and alloys which form thick oxide layers at ambient temperatures can tolerate also thick oxide layers before the hydrogen reaction is totally poisoned. The plateau length of the oxidation curves in Fig. 5.7 and the amount of oxygen needed for severe poisoning is about the same. This is demonstrated in Fig. 6.5 where the reaction probability r_{H_2} of the hydrogen absorption reaction is plotted, – here not as a function of the oxide layer thickness but as a function of the reaction probability for the oxidation reaction, r_{O_2}. The curves for the seven metals presented coincide within a relatively small scatter band. Heavy poisoning with r_{H_2} values below 10^{-5} is observed after the oxide layer thickness has reached a value where the reaction probability for oxygen absorption, r_{O_2} is reduced to values below 10^{-2}. This means according to the models described in Chap. 5 that the oxidation reaction has attained stage 2 where the Mott equilibrium potential breaks down and the electron flux through the oxide layer becomes rate determining.

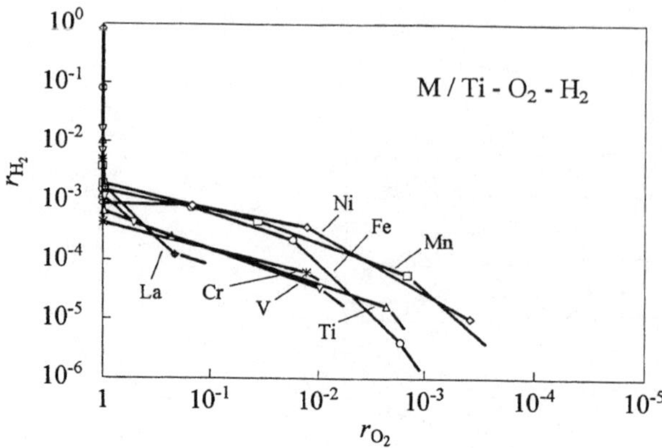

Fig. 6.5 Reaction probability for hydrogen, r_{H_2}, on various metal overlayers deposited on titanium films as a function of poisoning by the oxygen precoverage which is characterized by the oxygen reaction probability r_{O_2} [6.11]

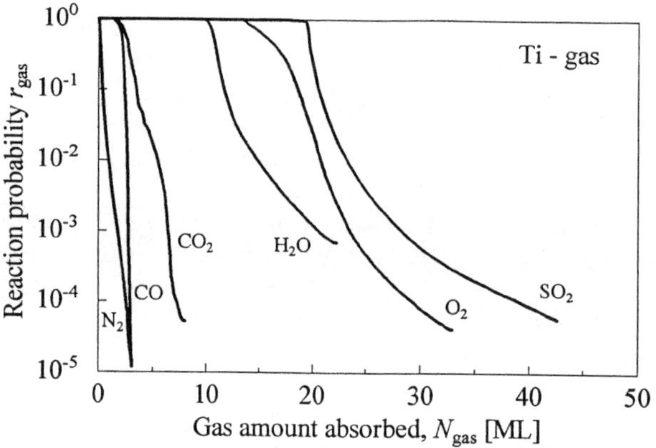

Fig. 6.6 Reaction probability of reactive gases, r_{gas}, measured in experiments on poisoning of hydrogen absorption of titanium films as a function of the amount of absorbed gas [6.11]

(ii) In another series of experiments poisoning by different reactive gases has been studied with titanium films as a convenient model substance [6.11]. Figure 6.6 shows the kinetics of contamination layers in a plot reaction probability of the gas molecule, r_{gas}, versus the amount of gas absorbed, N_{gas}, in monolayer equivalents. These curves are of the same type as those shown for oxygen absorption of metal films in Fig. 5.7. The layer thickness formed in the fast initial partial step of stage 1 is comparable to the length of the plateau at $r_{gas} = 1$. N$_2$ and CO form chemisorption surface layers which are only one or two monolayer thick if the film roughness factor of about two to four is considered. CO$_2$ produces slightly thicker layers, followed by H$_2$O, O$_2$, and finally SO$_2$. If the plateau length is correlated again with stage 1 of the low-temperature oxidation models then Mott-potential enhanced electron transfer should be possible through relatively thick oxide or sulfide layers formed in O$_2$, H$_2$O and H$_2$S. Nitrogen and carbon containing gases form only chemisorption layers in the monolayer range. Since the car-

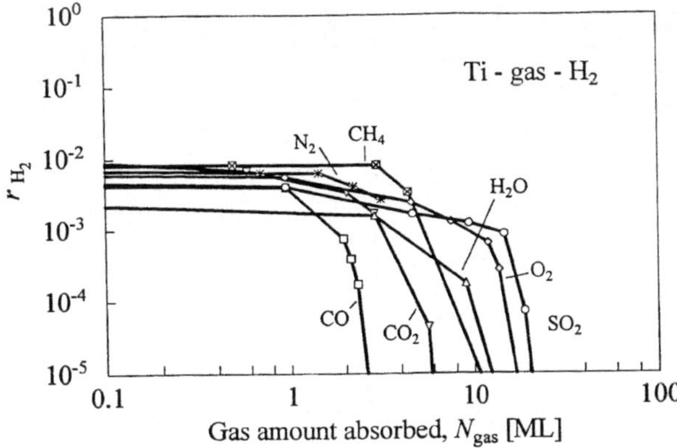

Fig. 6.7 Reaction probability of hydrogen, r_{H_2}, on titanium films poisoned by various gases as a function of the amount of absorbed impurity gas, H_{gas} [6.11]

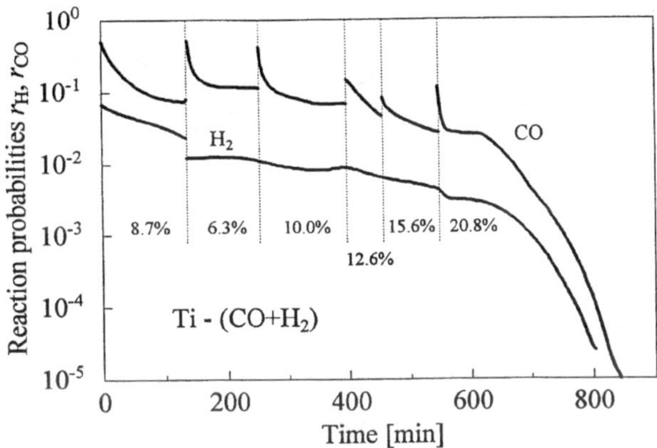

Fig. 6.8 Reaction probability of hydrogen and CO on a titanium film in H_2/CO gas mixtures with stepwise increased CO content [6.12]

bides and nitrides of titanium are metallic, the surface charges are balanced by mobile electrons in the bulk within a very short distance. A double layer is formed on the surface rather than a Mott equilibrium potential across a relatively thick semiconducting surface layer.

If the hydrogen absorption rate after poisoning by various gases is measured in a plot reaction probability r_{H_2} versus poison-layer thickness, given by the amount of absorbed poisoning gas, N_{gas}, then we realize again that severe poisoning is observed only at gas coverages beyond the plateau length in Fig. 6.6 where stage one in the oxidation models ends (Fig. 6.7). The layer thickness required for severe poisoning is given by the sequence CO, CO_2, H_2O, O_2, SO_2. N_2 does not poison the hydriding reaction remarkably.

Experiments performed with titanium films in mixtures of hydrogen gas with CO, N_2, H_2O and O_2, respectively, show the same trend [6.12]. The hydrogen reaction probability is reduced after the rate of the poisoning reaction has been

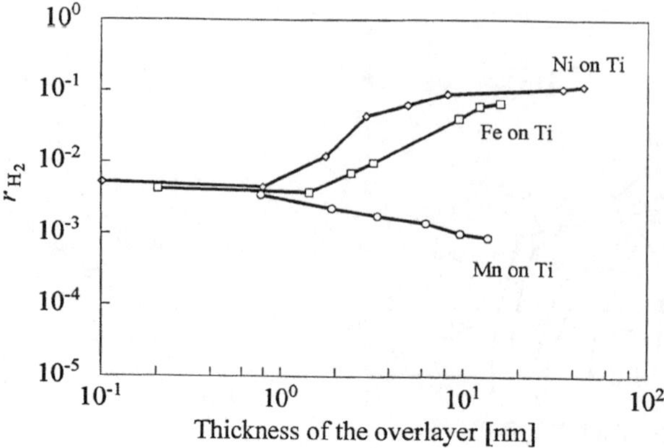

Fig. 6.9 Reaction probability of hydrogen as a function of the thickness of nickel, iron, and manganese overlayers on titanium films [6.13]

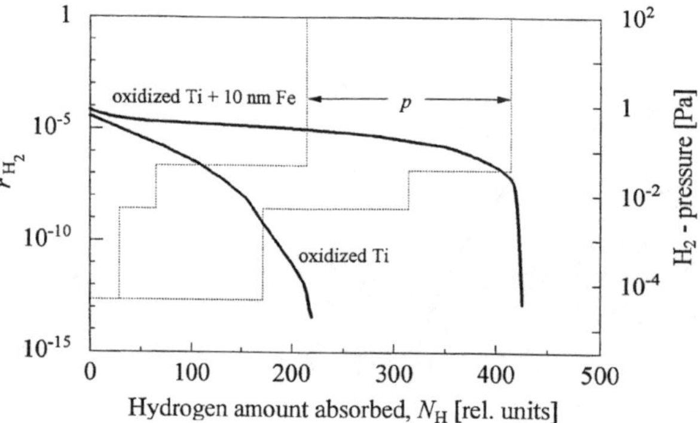

Fig. 6.10 Reaction probability of hydrogen on an oxygen poisoned titanium film without and after subsequent deposition of an iron overlayer [6.7]

strongly reduced. This is shown in Fig. 6.8 where r_{H_2} and r_{co} have been measured in H_2/CO mixtures of various composition in a combined experiment with the same titanium film.

(iii) The catalytic effects caused by surface layers of palladium, nickel and iron could be observed also with film samples. Figure 6.9 shows the enhancement of the hydrogen reaction rate of a clean titanium film covered by increasing amounts of nickel and iron [6.13]. The reaction probability r_H increases with the thickness of the overlayer. Only a metallic overlayer of about one nm thickness or metallic islands on the surface of equivalent thickness provide full efficiency and not the individual adsorbed metal atom. This demonstrates again that dissociation on the hydride overlayer is rate determining and not diffusion through the hydride layer. A manganese overlayer impedes the reaction rate slightly.

(iv) In another experiment performed with a quartz-crystal microbalance it could be demonstrated that an oxidized and severely poisoned titanium film is activated again by iron overlayers on the oxide layer [6.7]. This is shown in Fig. 6.10 by the two curves reaction probability r_{H_2} versus amount of absorbed hydrogen of a film oxidized by air exposure. The lower curve shows the first run with low r_{H_2} values in the range 10^{-7} after a few monolayers H_2 had been absorbed. The same film was then coated with 10 nm iron. The reaction probability r_{H_2}, is increased by almost ten orders of magnitude and relatively fast hydrogen absorption with a macroscopic tolerable rate of $10^{-5} < r_{H_2} < 10^{-7}$ was measured again over a long exposure period. This behavior, found also in other experiments with film samples, demonstrates that dissociation of hydrogen molecules on the surface of the oxide layer is much slower than permeation of hydrogen atoms through the oxide layer.

6.1.3 Powder Samples

Hydrogen storage materials of the prototype families FeTi, $LaNi_5$ or Mg(Ni) [6.16] are powders with particle sizes between 5 and 30 μm. The plateau pressure at the temperature of application (0–500 °C) is one to five bar. Poisoning of the absorption/desorption cycles is a very common but unpleasant experience and well known by experts working in the field. Technological investigations and observations in practical application demonstrate that additions of CO, O_2, H_2O, H_2S, and SO_2 to the hydrogen atmosphere act as severe poisons whereas N_2 and CO_2 and some of the hydrocarbons are less dangerous additives [6.17, 18]. Activation of powder samples is accomplished by several loading/unloading cycles in high-purity hydrogen gas, sometimes at elevated temperature. The improvement of the kinetics is mainly caused by cracking of the powder particles accompanied by reduction of the particle size and/or spalling of the contamination layers because of the lattice expansion during hydride formation. These mechanical effects create new, unpoisoned surfaces and it is difficult to separate this effect in an experiment from the recovering of poisoned original surfaces.

Recent studies on poisoning of $LaNi_5$ and $LaNi_{4.7}Al_{03}$ powders yield quantitative information on the reduction of the hydrogen absorption rate as a function of the amount of absorbed poisoning gases [6.19, 20]. This was accomplished by exposure of activated powder samples to known volumes of reactive gases and, in another series of experiments, by addition of reactive gases to the hydrogen gas atmosphere. In Fig. 6.11 experiments with stepwise increased CO precoverage of the particle surface are shown. Severe poisoning with a reduction of the absorption rate, v normalized to the rate of unpoisoned powders, v_0, by several orders of magnitude is also observed here if about one monolayer CO is absorbed. Figure 6.12 presents v/v^0 curves for loading/unloading cycles as a function of the number of cycles. A drastic reduction of the reaction rate is clearly correlated with the impurity content of the hydrogen gas. The CO impurity is accumulated on the powder samples until about one monolayer is reached and then the reaction rate is strongly decreased.

Figure 6.13 shows a compilation of experiments with five different gases as possible contaminants in hydrogen. Again, CO is able to poison the hydrogen absorption rate by a monolayer coverage. H_2S layers are thicker, O_2 layers demonstrate a more complex behavior and CO_2 seems to cause no real poisoning, only retardation of the hydrogen absorption rate by about one order of magnitude. N_2 does not show any remarkable effect at all. In experiments performed with mixtures of powder of storage material and Pd catalysers it could be demonstrated that poisoning can be reduced or partially avoided [6.13].

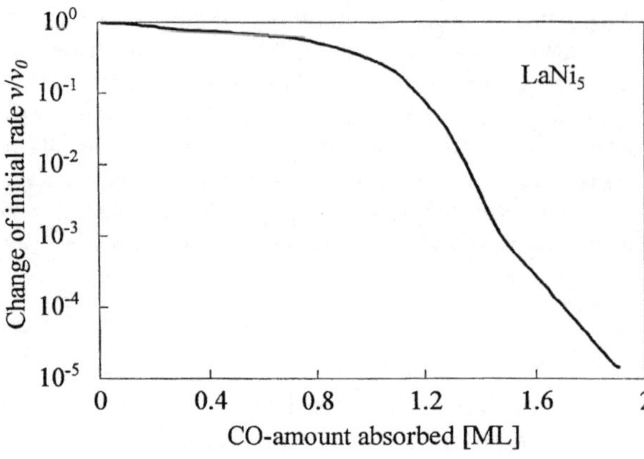

Fig. 6.11 Hydrogen absorption rate of LaNi$_5$ powder samples as a function of surface precoverage with carbon monoxide [6.19]

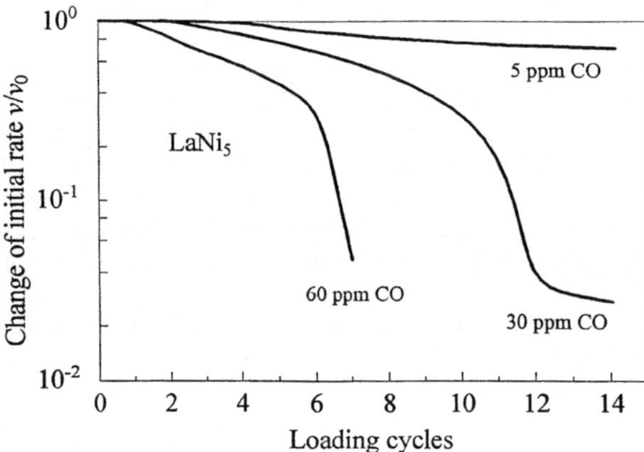

Fig. 6.12 Hydrogen absorption rate of LaNi$_5$ powder samples in H$_2$/CO mixtures as a function of loading/unloading cycles [6.19]

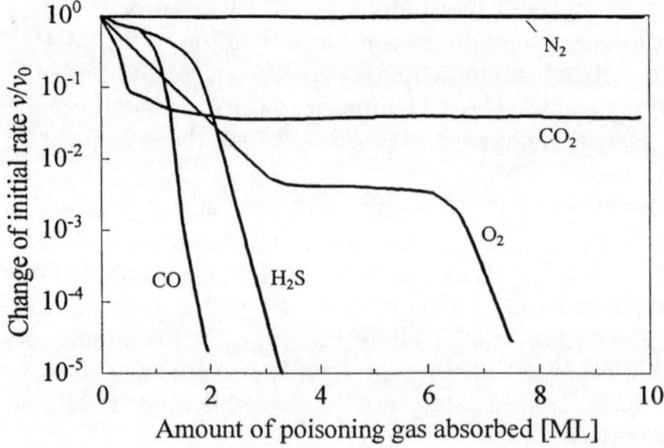

Fig. 6.13 Hydrogen absorption rate of LaNi$_5$ powder samples as a function of surface coverage with various gases[6.19]

For reasons mentioned above investigations on poisoning of powder samples suffer from problems caused by the undefined shape of the samples. Larger scatter of data is the normal situation and depends on specific treatments of the samples before hydrogen exposure and other details of the experimental procedure. Therefore, quantitative comparison with the results obtained at low pressures with wire and film samples is difficult. Nevertheless, the general trends observed look very similar.

6.1.4 General Trends

The examples presented here and many experimental results published by other authors reveal trends which seem to be now well settled as a general behavior typical for many metal hydrogen systems. They can be summarized by the following statements:

– Dissociation of the hydrogen molecule on the surface of a sample is a very important reaction partial step and most often rate determining at ambient or low temperature, especially with powder or film samples.
– On clean metal surfaces this process is very fast. On hydrides it can become slower with titanium by two orders of magnitude at ambient temperature. But this yields still a very high reaction rate at atmospheric pressure.
– For oxide layers and other contamination layers a critical layer thickness value exists. Below this value the reaction rate is reduced remarkably but it remains still high with reaction probabilities higher than about 10^{-4}. Above this critical thickness value a drastic reduction of the hydrogen absorption rate occurs and frequently hydrogen absorption is prevented quantitatively. This critical thickness can roughly be defined as the thickness where the reaction probability of the contamination layer formation has been reduced to $r_{gas} < 10^{-2}$.
– The high initial reaction rate in low-temperature oxidation depends on the existence of the Mott equilibrium potential or, in other words, on the presence of electrons on the surface for the formation of negatively charged surface species. It looks like that dissociation of hydrogen and of other physisorbed gas molecules is enhanced by electrons in surface states which are filled with electrons from the conduction band in the bulk. If metallic nitrides or carbides are formed as chemisorption layers, free electrons are generally available on the sample surface and no poisoning is observed. However, if the electrons from the bulk of the metal phase have problems to penetrate a semiconducting contamination layer, for example, an oxide scale, this condition fails. Then, the rate determining step for the hydrogen reaction is H_2 dissociation on the surface. It depends obviously in a similar way on the electron transfer through the contamination layer as the oxidation rate in the models on low-temperature oxidation.
– The catalytic effects of some more noble metals which are not oxidized in a H_2 atmosphere for thermochemical reasons, for example palladium, nickel, iron and others, can provide, too, the electrons needed for lowering the activation energy of H_2 dissociation on a passive oxide surface. After dissociation either a neutral hydrogen atom is formed or an ion plus an electron. In total an uncharged hydrogen atom has to penetrate the oxide scale to form a metal hydride. A Mott equilibrium potential would impede the flux of protons to the metal or to the hydride phase. Therefore, and for other reasons the presence of charged hydrogen species seems to be an unlikely assumption in model calculations.

6.2 Stability of Oxide Layers at Elevated Temperatures

The hydrogen absorption rate of reactive metals is impeded strongly at ambient temperatures by oxide surface layers or by other contamination films. This has been demonstrated by the examples presented in the preceding sections. However, at temperatures above 300° to 500 °C hydrogen is normally absorbed with perceptible rates comparable to the kinetics observed after an UHV heat treatment or of samples activated by other methods, for example, by deposition of a palladium layer as a catalyst. This indicates that either the dissociation of hydrogen molecules on the oxide surface and penetration of the oxide layer has become much faster or that the oxide layers have been removed. In this section it will be demonstrated by a semi-quantitative estimate that dissolution of oxide layers at elevated temperatures is a very likely process [6.21]. It is assumed that the thickness of the oxide layer is determined by the competition between the layer growth mechanism and the dissolution of oxygen atoms from the oxide layer in the bulk of the sample. The layer thickness is calculated as a function of time starting with an initial thickness of the layer. The simulated curves show whether growth or disintegration of an oxide layer is expected at given temperature and pressure.

6.2.1 Structure of the Model

From the experimental results and oxidation models discussed in previous chapters the following simplified behavior for oxide layer formation on a clean surface of reactive metals can be assumed (Fig. 5.7) :

(i) In the initial, very fast stage 1 more than the equivalent of one monolayer of oxygen is absorbed with a reaction probability r of one. Because of the extremely high reaction rate thermal activation can be disregarded.

(ii) In the second stage the reaction probability r is reduced either exponentially or inverse exponentially with increasing amounts N of absorbed oxygen atoms. The temperature dependence of r in the final region shall be represented by an activation energy A which approaches a constant value for large values of N.

(iii) The absorption rate in the first stage is controlled by transport in the gas phase and, thus, proportional to pressure. If in the final stage the reaction rate is surface controlled then the reaction probability r is again pressure independent and $v \sim p_{O_2}$ also holds here.

The main features of such a reaction mechanism can be simulated by the following simple equation for the layer growth:

$$v_1 = \frac{dl}{dt} = K_1 \cdot p_{O_2} \cdot \exp\left(-\frac{A}{kT} \cdot \frac{k_1 N}{1 + k_1 N}\right) \tag{6.1}$$

with l layer thickness; A, K_1, k_1, and k constants and N amount of absorbed O_2 gas. For small N values the reaction rate is almost not dependent on temperature and by proper choice of K_1 the reaction probability r can be made unity. For large values of N the reaction becomes a thermally activated process of first order with respect to pressure.

At elevated temperatures the oxygen atoms become mobile and disintegration of the oxide layer can occur due to solution of oxygen atoms in the bulk. If the metal phase can be approached by a semi-infinite space, the diffusion process is described by the error function (Fig. 2.5) and the concentration gradient at the metal/oxide interface $x = 0$ can be approximated by $\nabla c = 0.6\,c_S/\sqrt{Dt}$ with c_S terminal oxygen solubility in the α- phase at the temperature T chosen. Inserting the equilibrium constant for the solubility and the diffusion constant of oxygen in the metal in Fick's law yields

$$-v_2 = -\frac{dl}{dt} = \frac{K_2}{\sqrt{t}} \cdot \exp\left(-\frac{A_2}{kT}\right). \tag{6.2}$$

A_2 is an energy term containing the diffusion activation energy and the heat of oxygen solution, t is time, and K_2 is a constant. In a first approximation it can be assumed that the thickness of oxide layers on reactive metals is given as the sum of the growth rate (6.1) , which is strongly pressure and thickness dependent; and of the rate of the disintegration process (6.2), which is strongly temperature dependent. This estimate does not describe precisely the nonactivated initial stage one of the absorption reaction and the reaction mechanism for thick layers at elevated temperatures after transition to the final stage 3 with diffusion control. There a change of the rate law from a p_{O_2} to a $\sqrt{p_{O_2}}$ dependence must take place. Nevertheless, it demonstrates nicely the main feature of the interactions between both opposite processes.

6.2.2 Results

In Fig. 6.14 three groups of oxide thickness curves are shown for 300, 600, and 800 K. The constants in (6.1, 2) are compatible with experimental data for the Ta–O system. The solid lines represent the oxide layer thickness in ML O_2 for initially gas-free tantalum surfaces at different O_2 pressures and the dotted lines the change

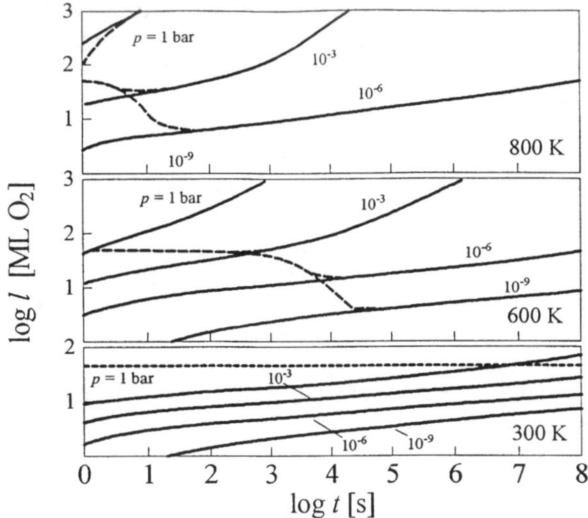

Fig. 6.14 Simulated curves for the oxygen layer thickness l on tantalum as a function of temperature and pressure. Solid lines, sample oxide free at $t = 0$; dotted lines, oxide layer 50 ML thick at $t = 0$

of an initial layer with a thickness of 50 ML O_2. At 300, 600, and 800 K thick oxide layers grow on gas-free surfaces within a short time in addition to the gas amount absorbed spontaneously during the initial fast reaction stage one (Fig.5.7). If it is assumed that the samples are covered at the beginning of the experiment by a thick oxide layer equivalent to 50 ML O_2 then the dotted curves are obtained. At 300 K no changes in the thickness growth curves occur in one year below one bar O_2. At 600 and 800 K the thickness curves are given first by the pressure independent disintegration curve and then, after the intersection point, by the growth curves for the pressure values chosen. This means that after a certain period of time, hours at 600 K and minutes at 800 K, the layer thickness attains a value which depends no longer on the initial layer thickness but only on pressure and temperature. The relative change of the thickness with time is then slow. At 800 K and 10^{-6} mbar O_2 the disintegration rate is faster than the growth rate and oxide layers are no longer stable under these conditions in our model.

6.2.3 Conclusions

The estimates of the model agree well with experimental observations on the kinetics of hydrogen absorption of tantalum and niobium wires in high-vacuum systems. The results can be considered as typical examples for the behavior of reactive metals covered at ambient temperature with a stable oxide skin. This yields the following view of the mechanism of H_2 absorption at low and elevated temperatures in high vacuum: Oxide layers are not stable on metal surfaces at temperatures higher than 800 K and oxygen pressures below 10^{-9} bar since bulk diffusion is faster than the O_2 transport in the gas phase. This result is not dependent on specific assumptions on the layer growth mechanism as long as the decomposition reaction at the oxide/metal interface is fast. For most metals data on the terminal solubility and the diffusion of oxygen atoms in the metal phase are available. For example they show that oxide layers on tantalum cannot be removed below 500 K by annealing less than one day even under extreme UHV conditions. Between 600 and 800 K and $p_{O_2} > 10^{-9}$ bar the thickness of the layer is slowly growing. After an incubation time between one hour and a few seconds it depends no longer on the pretreatment of the sample but only on pressure and, depending on the reaction mechanism, on temperature. The absolute value of the layer thickness in the region after the incubation period is the same as for initially uncovered samples. The pressure and temperature dependence of the growth rate v_l can deviate from the simple relations used here and cause changes in the shape of the growth curves in Fig. 6.14 but this does not affect the general trends of the phenomena observed. Since the transition in the curves from the initial branch of layer disintegration to the later branch of layer growth is rather abrupt, further simplification of the theoretical considerations is possible. If the oxidation curves for a specific system are plotted as a function of p_{O_2} one has only to superimpose the disintegration curves for the temperature chosen which depend only on the initial thickness of the oxide layer. By such an estimate semiquantitative information can be obtained on the surface state of reactive metals during vacuum annealing treatments at elevated temperatures. Many poisoning phenomena observed during hydrogen absorption and desorption of metallic samples or the effects of annealing treatments on the kinetics of hydrogen storage materials can be understood as a result of the competition between oxide layer growth and its dissolution in the metal phase.

6.3 Surface Layer of Constant Thickness

In this section problems will be considered which are similar to the problems of hydride scale formation on metals discussed in Sects. 4.3, 4 where systems with growing surface layers have been treated in the limit of one rate determining step and in the advanced model without this restriction. The present model describes the kinetics of hydrogen absorption or desorption of a metal phase which forms only a solid solution. But this reaction shall be impeded by a surface layer of constant thickness, for example by an oxide skin. The reaction mechanism is composed of the same partial steps as discussed in Sect.4.3.1:

1) Gas phase transport of H_2 molecules to the coating surface and physisorption.
2) Dissociation of adsorbed H_2 molecules and chemisorption of H atoms on the coating.
3) Transition of H atoms from the surface to the bulk of the coating.
4) Diffusion of H atoms through the coating.
5) Transition of H atoms to the metal at the coating/metal interface.
6) Diffusion inside the metal.

In the degassing process these partial steps are passed in reverse sequence. The theoretical treatment of this model is in between the methods discussed in Chap. 4. Instead of only one rate determining step two of them and their interactions are studied. For this problem a closed mathematical solution can be given which provides a more direct insight in linking two partial steps of a complex reaction than the computer simulation using the numerical methods discussed in Sect. 4.4.

The thickness of a surface coating on a metal sample can be considered constant if an oxide reaction has attained the stages 2 or 3 in the models on low-temperature oxidation (Chap. 5)where the growth rate is very small. In the limits for pure dissociation control or pure interface control the results are the same as discussed in Sects. 4.2.4, 6 and 4.3.2, 3. If diffusion of hydrogen atoms through an oxide scale is rate determining, the permeation flux is given by the equilibrium activities of hydrogen on both sides of the coating. It is this reduction of the diffusion problem to a permeation problem which enables mathematical solutions in closed form for models with two rate determining steps.

A model with two rate determining steps has been developed by *Nakamura* [6.22] for systems where permeation through a coating of constant thickness and hydrogen diffusion in solid solution of a metal sheet are combined. It provides solutions for a system where the H_2 dissociation or other reactions on the coating surface and all other interface processes are defined to be fast reaction steps. These assumptions can be realistic for reactive metals coated by noble metal layers as catalysts [6.15]. The structure of this model is discussed below in Sect. 6.3.4.

The other model presented in Sects. 6.3.1–3 with two rate determining steps investigates the interrelations between thermally activated H_2 dissociation on the coating surface and the permeation of hydrogen atoms through the coating. Diffusion in the α- metal phase and transition of hydrogen atoms through the metal/coating interface are assumed to be fast reactions [6.23]. This model is of specific interest for the interpretation of hydrogen absorption and desorption kinetics of reactive metals coated by oxide skins at ambient temperatures. If permeation of hydrogen through a surface layer of constant thickness is defined to be also a *sur-*

face reaction and diffusion in the α-solution phase is assumed to be fast, then both *surface processes* lead to a constant flux to the bulk during the initial stage of the absorption reaction far from equilibrium. The time law remains, therefore, always linear, $c \propto t$. However, the pressure dependence of the absorption rate is changing from $v \propto p$ for dissociation control to $v \propto p^{1/2}$ for permeation control. The behavior of the Nakamura model (Sect. 6.3.4) is exactly the opposite. There the pressure dependence of the absorption rate is always $v \propto p^{1/2}$ and the time law is changing from $c \propto t$ for permeation control to $c \propto t^{1/2}$ for control by bulk diffusion in the α-phase.

In this section, the main results of the model with dissociation and permeation control will be presented at first since it considers the kinetics of hydrogen absorption as well as of desorption. The results are presented as simulated absorption curves in plots concentration versus time for various exposure pressures.

6.3.1 Absorption, Dissociation and Permeation Control

The H_2 absorption reaction shall proceed according to the following reaction scheme (Fig. 6.15): H_2 molecules approach to the surface and are adsorbed in the shallow molecular physisorption state, ph, on the coating surface (flux J_{gas}). Then, the H_2 molecules are dissociated on the surface and reach the chemisorbed state, ch, by an activated reaction step ($J_{ph,ch}$). From this state the hydrogen atoms can recombine again and are desorbed ($J_{ch,ph}$) or they are transferred to position 0 inside the coating. Between ch and 0 the segregation equilibrium for hydrogen atoms is established. Hydrogen atoms diffuse inside the coating to position B (J_{diff}). Between B and M the segregation equilibrium for hydrogen atoms is established, too, and inside the metal no hydrogen concentration gradient exists due to a fast diffusion flux J_m.

Definitions and interrelations between individual fluxes
a) Flux in the gas phase:

$$J_{gas} = k_g p. \tag{6.3}$$

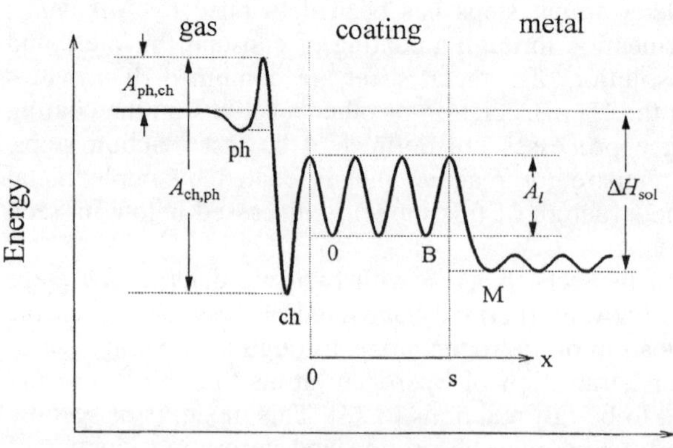

Fig. 6.15 Energy reaction-path diagram (schematic)

If the number of absorbed hydrogen atoms is small compared with the number of impinging H_2 molecules (reaction probability $r \ll 1$) then the coverage $\theta_{H_2(phys)}$ of the surface with H_2 molecules is in equilibrium with the gas phase,

$$\theta_{H_2(phys)} = K_p p = (K_p/k_g)J_{gas} \tag{6.4}$$

b) Flux to the chemisorbed state and dissociation of H_2 :

$$J_{ph,ch} = k'_c \theta_{H_2(phys)} = k'_c K_p p = k_c p. \tag{6.5}$$

c) Flux from the chemisorbed state and recombination of hydrogen atoms:

$$J_{ch,ph} = k'_{-c} \theta_{H(chem)} = k_{-c} a^2. \tag{6.6}$$

a_0 is the activity of hydrogen in the oxide subsurface in equilibrium with the chemisorption coverage.

d) Permeation flux through the coating layer:

$$J_{diff} = k_1(a_0 - a_B) \tag{6.7}$$

with $k_1 \approx D/l$; l being the coating thickness.

e) Parameter P:
Calculations are facilitated if the parameter P is introduced which depends only on temperature and the coating thickness, l,

$$P = k_c p^{1/2}[k_1 K_S]^{-1} = k_c p[k_1 a_e]^{-1} = J_{ph,ch}/J_{diff,max}. \tag{6.8}$$

$J_{diff,max}$ is the maximum permeation flux for $a_0 = a_e$; $a_B = 0$ (6.7). The constant activity value a_e is the hydrogen activity at equilibrium with the gas phase and $K_S = a_e/p^{1/2}$ the Sieverts' constant (Sect. 2.1).

f) Steady state flux:

$$J = J_{diff} = J_{ph,ch} - J_{ch,ph}. \tag{6.9}$$

This assumption means that concentration changes at ph, at ch, and in the layer consume only a small part of the overall hydrogen flux (quasi-steady state condition). Then for the fluxes in position ch the equation holds

$$J_{ph,ch} - J_{ch,ph} - J_{diff} = 0. \tag{6.10}$$

By inserting (6.5–7) in (6.9) the hydrogen activity a_0 at the coating surface is given by

$$a_0 = (k_1/k_{-c})\left\{\left[1/4 + (k_{-c}/k_1)a_B + (k_c k_{-c}/k_1^2)p\right]^{1/2} - 1/2\right\}. \tag{6.11}$$

With this expression and (6.7) equation (6.9) becomes

$$J = (k_1^2/k_{-c})\left\{\left[1/4 + (k_{-c}/k_1)a_B + (k_c k_{-c}/k_1^2)p\right]^{1/2} \right.$$
$$\left. - [1/2 + (k_{-c}/k_1)a_B]\right\}. \tag{6.12}$$

p in (6.12) can be expressed by the equilibrium activity

$$a_e = p^{1/2} \cdot K_S. \tag{6.13}$$

In equilibrium $J_{ph,ch} = J_{ch,ph}$ holds and (6.5, .6, 13) yield

$$K_S^2 = k_c/k_{-c} = a_e^2/p. \tag{6.14}$$

With (6.8) equation (6.12) can be written as

$$J = (J_{ph,ch}/p^2)\left\{\left[1/4 + P\,a/a_e + p^2\right]^{1/2} - (1/2 + P\,a/a_e)\right\}. \tag{6.15}$$

g) Reaction probability, r.
The reaction rate of H_2 absorption can also be represented by the ratio of H_2 molecules being absorbed to H_2 molecules hitting the surface.

$$r = J/J_{gas} \tag{6.16}$$

and the maximum reaction probability r^0 in terms of the maximum flux J_c is

$$r^0 = J_{ph,ch}/J_g = k_c/k_g. \tag{6.17}$$

With (6.15, 16) equation (6.17) becomes

$$r/r^0 = J/J_{ph,ch} = (1/P^2)\left\{\left[1/4 + P\,a/a_e + p^2\right]^{1/2} - (1/2 + P\,a/a_e)\right\}. \tag{6.18}$$

Equations for the steady state flux J and the reaction probability r
J and r are defined by (6.15, 18) as functions of p, T, the parameter P, and a term for the activity a/a_e in the metal normalized by the equilibrium activity a_e. If Sieverts' law for ideal dilute solutions is valid

$$a/a_e = x/x_e \tag{6.19}$$

holds. The parameter P is defined as

$$P = k_c p [k_1 a_e]^{-1} = J_{ph,ch}/J_{diff,max} \tag{6.20}$$

or the ratio of the chemisorption flux $J_{ph,ch}$ and the maximum permeation flux $J_{diff,max}$ established in the coating if $a_0 = a_e$ and $a_B = 0$ holds (initial stage of permeation control). Thus, for $P \ll 1$ the reaction is surface controlled ($J_{ph,ch} \ll J_{diff,max}$) and for $P \gg 1$ it is permeation controlled ($J_{diff,max} \ll J_{ph,ch}$). In these two limits simpler equations are obtained for J and r. Further limits are given for the initial stage, $a/a_e \ll 1$, and for the final stage, $\Delta a = (a_e - a) \ll a_e$, of the reaction. Equations for these limiting cases are compiled in [6.23].

Time Law of Absorption
The concentration change in the bulk, dx/dt, is related to the steady state flux by

$$J = V/S \cdot dx/dt = V/S\gamma_m \cdot da/dt \tag{6.21}$$

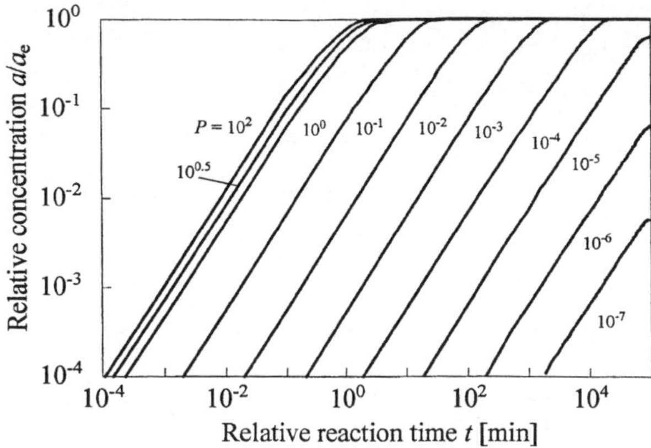

Fig. 6.16 Relative concentration a/a_e vs. reaction time for various values of the parameter P (6.20)

with V volume and S surface area of the sample and γ_m activity coefficient. Inserting (6.14, 15) in (6.21) and subsequent integration finally gives an equation

$$(S\gamma_m k_1/V) \cdot t = F(a/a_e, P) \tag{6.22}$$

which is lengthy and given in [6.23]. It represents the change of the relative activity a/a_e as a function of time with P as a parameter depending only on p, T and l (6.8, 20). In Fig. 6.16 simulated absorption curves are shown for typical values of P. In the plot log a/a_e vs. log t the curves have a linear initial branch followed by the region where the equilibrium activity $a/a_e = 1$ is approached. Equations for limiting cases of (6.22) are also given in [6.23].

Pressure dependence of the initial absorption rate
The initial absorption rate v^0 is defined as

$$v^0 = dx/dt\big|_{a=0} = (S/V)J_{a=0} \tag{6.23}$$

or with (6.15)

$$v^0 = (S/V)(k_c p/P^2)\left[(1/4 + P^2)^{1/2} - 1/2\right]. \tag{6.24}$$

In experiments usually the exponent n of the empirical formula $v = k(T, x)p^n$ is determined. It is defined in the model by the equation

$$n = d\ln v^0/d\ln p = (p/v^0)(dv^0(dp) = F(P) \tag{6.25}$$

or

$$n = F(P) = (P^2/2)\left[(1/4 + P^2) - 1/2(1/4 + P^2)^{1/2}\right]^{-1} \tag{6.26}$$

with the limits: Surface control

$$P \ll 1 : n = 1 \qquad \text{and} \qquad v^0 \propto p$$

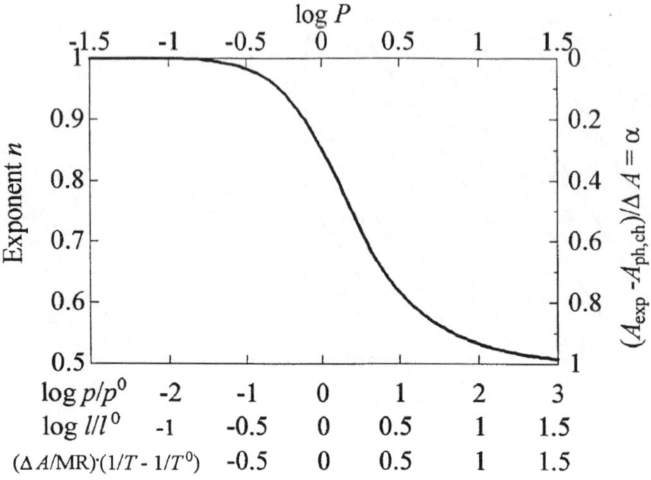

Fig. 6.17 Function $n = F(P)$, (6.28), and $(A_{\mathrm{exp}} - A_{\mathrm{ph,ch}})/\Delta A = \alpha(P)$ (6.34), vs. $\log P$ or $\log p/p^0$; $\log l/l^0$ or $\Delta A/MR \cdot (1/T - 1/T^0)$, respectively

and permeation control

$$P \gg 1 : n = 1/2 \quad \text{and} \quad v^0 \propto p^{1/2}.$$

Figure 6.17 shows the dependence of the exponent n as a function of the parameter P in a plot $F(P)$ vs. $\log P$. P depends on the coating thickness, l, the H_2 pressure, p, and the temperature, T.

$$P = lp^{1/2}(k_c^0/k_1^0 K_S^0)\exp(\Delta A/RT) \tag{6.27}$$

with $\Delta A = (A_. + \Delta H_{\mathrm{sol}}) - A_{\mathrm{ph,ch}} \cdot A_l$. A_1 is the activation energy of hydrogen diffusion in the coating, ΔH_{sol} the heat of hydrogen solution, and $A_{\mathrm{ph,ch}}$ the activation energy of chemisorption. The dependence of n on the experimental parameters l, p, and T can be demonstrated by introducing the definitions

$$\log P|_{p,T} = \log l/l^0, \qquad\qquad P = 1 \quad \text{for} \quad l = l^0, \tag{6.28}$$

$$\log P|_{l,T} = 1/2 \log p/p^0, \qquad\qquad P = 1 \quad \text{for} \quad p = p^0,$$

$$\log P|_{l,p} = (\Delta A/MR)(1/T - 1/T^0); \qquad P = 1 \quad \text{for} \quad T = T^0.$$

The scales at the lower end of Fig. 6.17 are labeled according to the definitions of (6.28).

Temperature dependence of the initial absorption rate
The experimental activation energy A is defined by the Arrhenius relation

$$A_{\mathrm{exp}} = -R\, d\ln v^0/d(1/T). \tag{6.29}$$

(6.24, 27, 29) yield

$$A_{exp} = A_{ph,ch} + \Delta A \cdot 2[1 - F(P)]. \tag{6.30}$$

It can be demonstrated that the limits hold [6.23]

Surface control $(P \ll 1)$: $A_{exp} = A_{ph,ch}$;

Permeation control $(P \gg 1)$: $A_{exp} = A_l + \Delta H_{sol}$.

In Fig. 6.17 the variation of the activation energy A is on the right-hand scale presented in relative units $\alpha = (A_{exp} - A_{ph,ch})/\Delta A$. It shall be mentioned that for positive and for negative ΔA values A_{exp} is larger at low temperatures and decreases with rising temperature [6.23]. The temperature range of transition of the $\alpha(P)$ curves depends in a similar way on T_0 and ΔA as $F(P)$ (Fig.6.17).

6.3.2 Discussion of the Mechanism

In the model it is assumed that the absorption rate of hydrogen by a metal phase is determined by dissociation of H_2 molecules on the surface of a coating of constant thickness and/or permeation of hydrogen atoms through the coating. The results are absorption curves a/a_e vs. t and information on the dependence of the initial absorption rate $v^0(t = 0)$ on pressure and temperature. These data can be compared with experimental findings. If we wish to understand how the model works, we have to look at the values of the hydrogen activity as the driving force of kinetics. The hydrogen activity in the gas phase is given by the pressure. In the α-phase the activity is very low in the beginning and approaches as a function of time the value of the gas phase in the course of the simulation experiment. The aim of the model calculation is to get this function or the time law of the reaction. A very important hidden variable of the model is the activity of hydrogen in the chemisorbed state. We have defined that it must be equal to the activity in the subsurface plane of the layer. Thus, a_{chem} and a_0 are virtually the same variable. The value of this variable lies between the constant hydrogen activity in the gas phase or in the physisorbed state and the slowly changing activity in the metal phase. It is simultaneously a variable of the rate equations of the two rate determining partial steps, namely, of the dissociation backward reaction (6.6) and the permeation flux (6.7). The value of a_{chem} is determined by (6.9) so that the steady state condition is obeyed. The rest of the model is mathematics.

The equations obtained are characterized by the parameter P, which is dependent on the coating thickness, l, the H_2 pressure, p, and the temperature, T. The parameter P is defined as the ratio $J_{ph,ch}/J_{diff,max}$, which is the ratio of the chemisorption flux $J_{ph,ch}$ and the maximum permeation flux $J_{diff,max}$ under the specific conditions defined by an experiment. Consequently, the equations for dissociation control $(P \ll 1)$ and for permeation control $(P \gg 1)$ have to yield the same expressions as obtained by the simpler models for the limit of only one rate determining step. The mathematical treatment with results in closed form has the advantage that this condition can really be proved [6.23].

The most important result of the model is the calculation of the dependence of v^0 on the H_2 pressure according to the empirical relation

$$v^0 = p^n; \quad 0.5 < n < 1, \tag{6.31}$$

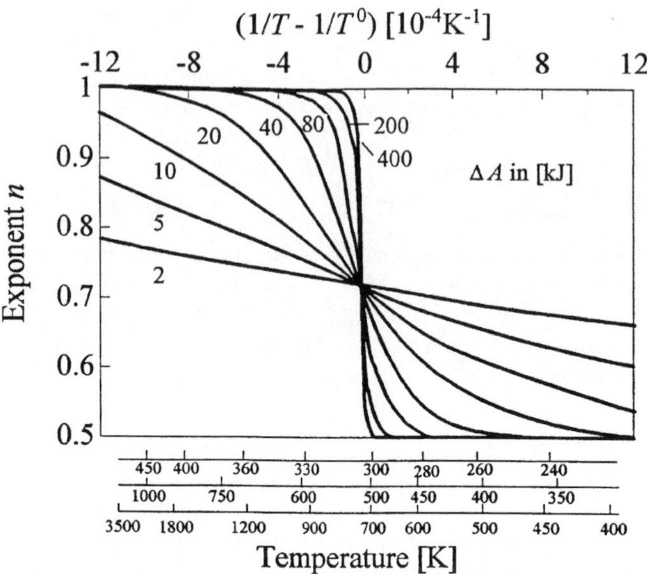

Fig. 6.18 *Function* n = F(P) vs.$(1/T - 1/T^0)$ or vs. *T* for various positive ΔA values

which can be determined by experiment. This demonstrates that one does not need to introduce specific mechanisms like pipe diffusion through cracks or migration of complex defects inside the coating to explain deviations from a $v^0 \sim p^{1/2}$ relation. The analysis of experimental data with respect to the present reaction model can proceed by the following steps: The parameter P is obtained from the exponent n of the pressure dependence of v^0 in Fig 6.17. This indicates whether the condition of one of the two limiting cases is fulfilled or that the experiment belongs to a run in the transition region $0.1 < P < 10$. If the equations for the transition region are applied, it can be checked by the dependence of the exponent n on p, T, and l whether or not the model is able to explain the experimental results.

From exact measurements of the exponent n of the pressure dependence of v^0 and of the activation energy A_{exp} and the determination of their dependence on p, l, and T further information on the energy terms used in the model can be obtained. The activation energy varies from $A_{\text{ph,ch}}$ to $(A_{\text{dl}} + \Delta H_{\text{sol}})$ if $|\Delta A/MRT|$ is changed by an amount of about two (Fig. 6.15). The extension of the transition temperature range for n and A_{exp} depends strongly on ΔA and T. This is demonstrated in Fig. 6.18 where transition curves of n vs. $1/T$ for different ΔA values are shown. Only for large ΔA and low T values the transition region is small. The ΔA values must be larger than 20 kJ to yield pronounced kinks in experimental curves $\log v^0$ vs. $1/T$ (Fig. 6.19).

The model can be checked in principle also by the shape of $\log a/a_{\text{e}}$ vs. $\log t$ curves. Figure 6.20 shows the different approach of the activity a/a_0 to the equilibrium value of 1 in the final stage of an absorption experiment. For surface controlled reactions ($P \ll 1$) the initially linear branch of the curve is more extended than for permeation controlled reactions ($P \gg 1$). However, small deviations in the shape of absorption curves in the final stage of an experiment cannot be easily measured and analyzed quantitatively.

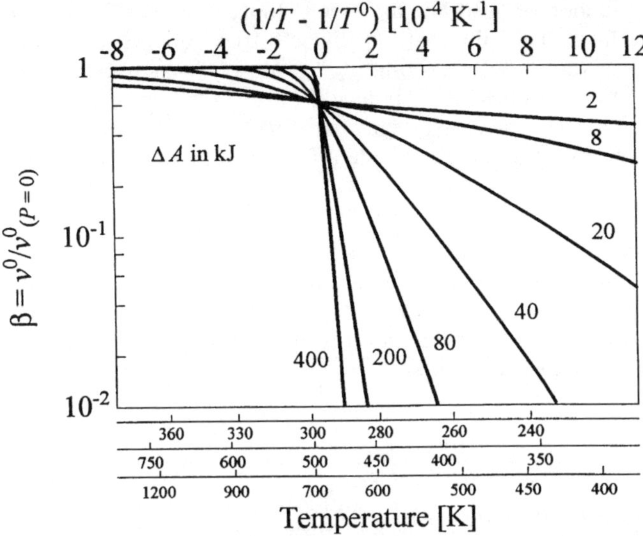

Fig. 6.19 Change of the initial absorption rate $\beta = v^0/v^0_{p=0}$ with temperature for different ΔA values. $v^0_{P=0}$ is the v^0 value for the limit $P \ll 1$

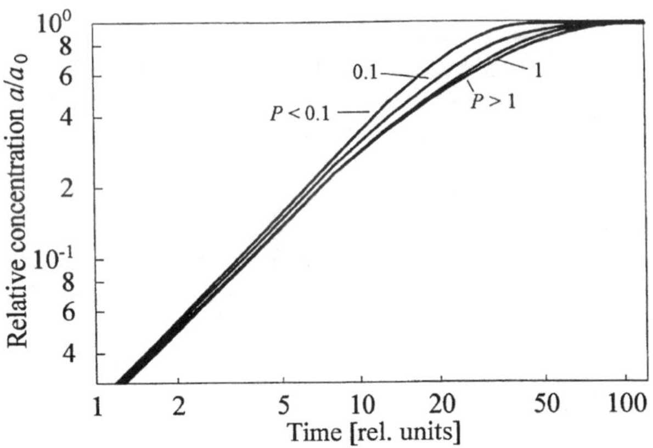

Fig. 6.20 Shape of log a/a_e vs. log t curves in the final stage of absorption experiments for different P values

6.3.3 Desorption, Surface and Permeation Controlled

In the desorption reaction hydrogen dissolved in a metal which is coated with an oxide scale has to pass the following steps in our model (Fig. 6.15): Diffusion of hydrogen atoms to the metal/coating interface and transition to position B in the coating are assumed to be fast partial steps. Then hydrogen atoms diffuse through the coating to position 0 (J_{diff}) which can be a slow partial step. Transition from 0 to the chemisorbed state at the surface shall be again a fast partial reaction. The transfer of hydrogen atoms from the chemisorbed state to the gas phase ($J_{\mathrm{ch,ph}}$) is accomplished by recombination of adsorbed hydrogen atoms and desorption of H_2 molecules from the surface. This thermally activated step can be slow, too. Furthermore, thermally activated chemisorption ($J_{ph,ch}$) can occur as backward reaction. H_2 molecules in the gas phase and molecular adsorbed hydrogen are assumed to be in equilibrium. The pressure dependent flux $J_{ph,ch}$ defines the final hydrogen equilibrium concentration in the sample after extended desorption.

Mathematical treatment of the model
The mathematical treatment of the model is very similar to the procedures applied for the absorption model [6.23] and will not be discussed here in detail. Only the general feature of some more important results will be presented and compared with curves for the absorption model. In addition to the definitions b), c) and d) mentioned above the following relations hold:

h) Parameter Q
Calculations are facilitated here also by introducing a new parameter $Q(T, l)$.

$$Q = k_{-c}/k_l \quad \text{or} \quad Q_a = k_{-c}a^2/k_l a = -J_{ch,ph}/J_{diff,max}. \tag{6.32}$$

$J_{diff,max}$ is the maximum permeation flux from B to 0 for $a_B = a$; $a_0 = 0$ (6.7). For $Qa \ll 1$ the reaction is surface controlled ($-J_{ch,ph} \ll J_{diff,max}$) and for $Qa \gg 1$ it is permeation controlled ($J_{diff,max} \ll J_{ch,ph}$). In these two limits again the simpler equations are obtained for the flux equations and time laws valid for the models with one rate-determining step [6.23]. Simpler formulas can be derived for the initial stage of the desorption process ($a = a_i \gg a_e$) and the final stage where the activity in the sample approaches the equilibrium value a_e. The final activity a_e is determined by the residual gas pressure in the receiver.

Typical results of the model are the curves in Fig. 6.21 which give the relative decrease of the hydrogen concentration in the sample a/a_i as a function of time for $a_e = 0$. The quantity Qa_i is used as a parameter which depends only on the initial concentration a_i, the layer thickness l, and the temperature. The curves for $Qa_i \gg 10^3$ coincide and the curves for $Qa_i < 1$ are shifted parallel to the $\log t$

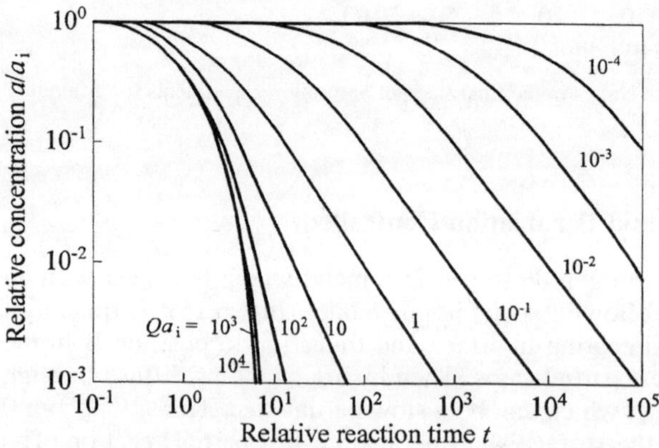

Fig. 6.21 Desorption curves $\log a/a_i$ vs. $\log t$ for different values of Qa_i and $a_e/a_i < 10^{-4}$

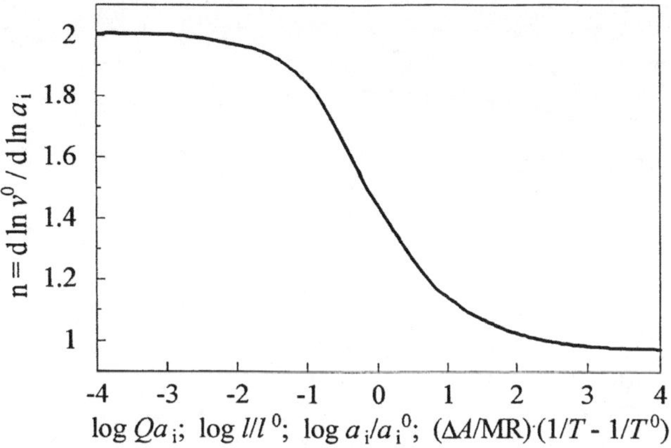

Fig. 6.22 Function n = d ln v^0 /d ln a_i vs. log Qa_i, log a_i/a_i^0; or $(\Delta A/MR)(1/T - 1/T^0)$ for $a_e/a_i \ll 1$

axis. The slopes of curves with large Qa_i values are steeper than for curves with small Qa_i values. The initial desorption rate is given by the expression

$$v^0 = dx/dt|_{a=a_i} = (S/V)J_i; \qquad J_i = J_{a=a_i}. \tag{6.33}$$

It has been demonstrated in Sect.4.2.5 that $J \propto a^2$ holds for the limit of surface control ($Qa_i \ll 1$) and $J \propto a$ for permeation control ($Qa_i \gg 1$). The exponent n in a^n is changing again smoothly from 1 to 2 as a function of the parameter Qa_i. This can be seen in Fig. 6.22. The scales are labeled by analogy to the definitions used in Fig. 6.17. The curves in Figs 6.21, 22 show that also for desorption reactions the transition from one limit to the other is not an abrupt one but requires normally parameter changes of more than one order of magnitude.

6.3.4 Absorption, Permeation and Diffusion Control

Based on the reaction scheme characterized by the energy diagram in Fig. 6.15 another model has been developed with two overlapping rate determining steps by *Nakamura* [6.22]. He combines slow permeation through a surface layer of constant thickness with diffusion in the bulk of sheet and wire samples. Dissociation and recombination of H_2 molecules on the layer surface shall be fast reaction partial steps. The mathematical treatment also provides here results in closed form. A parameter L is introduced which characterizes the ratio of the maximum diffusion fluxes in the bulk and in the coating:

$$L = \frac{(D_1 \cdot S_1)/l}{(D_m \cdot S_m)/d} \tag{6.34}$$

with D and S as the diffusion and Sieverts' constants in the layer and the metal phase and l and d the thickness of the layer and the sheet. The time law is given by an expression

$$a/a_e = 1 - \sum_{n=1}^{\infty} \frac{2L^2}{\beta_n^2(\beta_n^2 + L^2 + L)} \exp(-\beta_n^2 Dt/d^2). \tag{6.35}$$

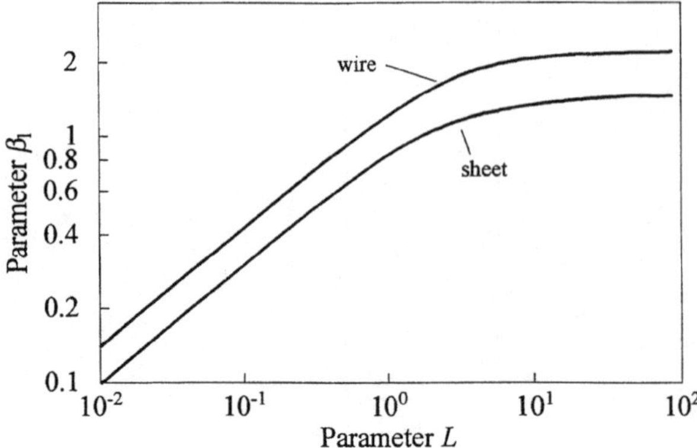

Fig. 6.23 Roots of the parameter $L = \beta_1 \tan \beta_1$ as a function of L

β is defined by the function $\beta_n \tan \beta_n = L$. Figure 6.23 shows the factor β_1 which appears in the most important first term of the sum as a function of L. The straight line on the left side for $L \ll 1$ describes the limit of permeation control and the plateau at the right side the region of bulk diffusion control. A very similar solution is obtained for wire samples. The curves demonstrate that also in this model transition from one limit to the other requires a ratio between the two partial diffusion fluxes considered of clearly more than one order of magnitude. Readers who are interested in further results and information offered by the model are referred to the original paper [6.22]. If this model is used one has to keep in mind that it simulates systems where diffusion of hydrogen in the bulk of the sample is not much faster than permeation through the coating or other reaction partial steps at the surface of the sample. This unusual situation can be realistic for metal samples coated with metal films which are not oxidized in a hydrogen atmosphere, for example with a palladium film.

6.4 Contamination Layers Growing During Exposure

If reactive metals with cleaned surfaces are exposed to a hydrogen gas atmosphere the oxygen or water molecules present as impurities will cover the surface again with a poisoning oxide scale. Then, two reactions are proceeding in parallel and affect each other as shown in Fig. 6.24. The mechanism of oxide scale growth on a metal hydride is almost the same as discussed in Chap. 5. The difference is the presence of hydrogen molecules and atoms on the oxide surface and of hydrogen atoms inside the oxide scale. After formation of a new oxide molecule or incorporation of a metal atom in the oxide scale at the oxide/hydride interface hydrogen atoms are released there. They can increase the hydrogen concentration in the hydride phase; they can migrate to the hydride/metal interface and form there new hydride molecules or they can be released back to the gas phase. The reactivity of oxygen in contact with metal atoms is much higher than the reactivity of hydrogen and, therefore, hydrogen present on the surface, on interfaces or as metal hydride is normally not able to affect the oxidation reaction remarkably. The ki-

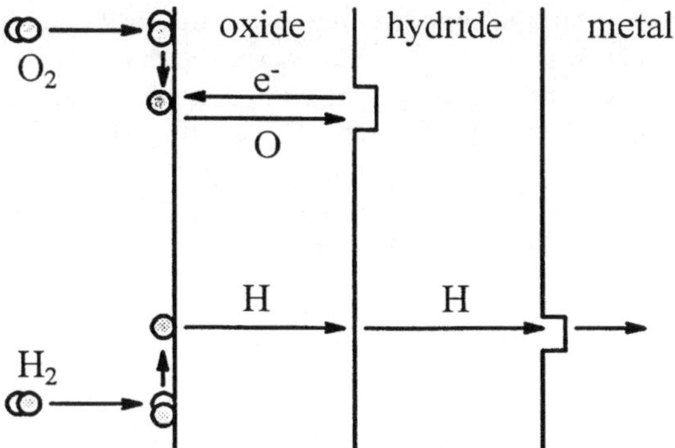

Fig. 6.24 Reaction mechanism of poisoning the hydride formation by oxide layers, schematic

netics of oxygen absorption of clean metal films and of films with stepwise changing oxygen and hydrogen exposures was found to be comparable [6.9, 10].

The situation for the hydrogen reaction is quite different. The kinetics is changed strongly if an oxide layer is present on a metal surface. It has already been discussed in the models of the previous sections that the sensitive process of the dissociation of hydrogen molecules is now performed at the oxide surface and hydrogen atoms have to migrate through the oxide scale and penetrate the oxide/hydride interface. After the subsurface site in the hydride layer has been reached the same reaction partial steps are passed as discussed in Sect. 4.3.1.

A consequent structure of a model on the reactions of a metal sample at ambient temperature in hydrogen/oxygen gas mixtures should include all reaction partial steps of both parallel reactions, the oxidation process (Sect. 5.3.1) as well as hydride formation (Sect. 4.3.1). In addition also interactions of hydrogen atoms, present on the surface and in the bulk, with the oxide phase have to be considered (Sect. 6.3). A treatment of such a mechanism in full length is complex and not well suited for a detailed analysis. Therefore, the discussion of the problem will be reduced here again to the approach of one rate determining step. Solutions of this kind of simple models are still useful since they can explain the general trends observed by experiments by reasonable approximations. For the development of such estimates, it is a big advantage that poisoning effects observed with hydrogen absorption and desorption reactions are very pronounced and the reaction rates are changed by many orders of magnitude. In practical application only the early stages of poisoning are of interest where the absorption rate is lowered by one to three orders of magnitude. Theoretical rate and time laws for the very slow kinetics of severely poisoned samples are normally not needed for the understanding of processes in real systems. The more important initial behavior can be simulated semiquantitatively by relative simple models if the highly complicated real reaction mechanism composed of more than ten partial steps is reduced to few basic processes responsible for the strong impediment of the overall reaction.

6.4.1 Poisoning by Chemisorption Layers

The hydrogen absorption of a metal sample is usually prevented if the surface is covered by a one monolayer thick chemisorption layer of carbon monoxide. This process can be simulated by a relative simple model which is based on the conception of co-adsorption and blocking of active sites required for the dissociation of H_2 molecules to form chemisorbed hydrogen atoms. It can be easily obtained by adaptation of the model with dissociation as the rate determining step. The empty reactive sites for hydrogen and CO chemisorption are then given by a term $(1 - \theta_{CO} - \theta_H)$ in (4.16). If the contamination gas is assumed to be chemisorbed reversibly then the steady state condition for hydrogen is represented as

$$\frac{d\,\theta_{H(chem)}}{dt} = 0 = k_{H+}\theta_{H_2}(1 - \theta_{CO} - \theta_{H(chem)})^2 - k_{H-}\theta_{H(chem)}^2 - J_{diff}. \tag{6.36}$$

For the contaminant CO the rate law can be written as

$$\frac{d\,\theta_{CO}}{dt} = k_{CO+}p_{CO}(1 - \theta_{CO} - \theta_{H(chem)}) - k_{CO-}\theta_{CO}. \tag{6.37}$$

Poisoning of the hydrogen reaction happens in this model if the CO adsorption reaction is irreversible or $k_{CO-} \ll k_{CO+} \cdot p_{CO}$ holds. Since chemisorbed hydrogen atoms are replaced by chemisorbed CO molecules, $\theta_{H(chem)}$ can be neglected for the moment. Then θ_{CO} obeys the time law

$$\theta_{CO} = 1 - \exp(-k_{CO+}p_{CO}t). \tag{6.38}$$

If the dissociation process for H_2 molecules is assumed to be rate determining then only the forward reaction rate has to be considered at the beginning of an experiment:

$$v = v^0 \cdot p_{H_2}[\exp(-k_{CO+}p_{CO}t) - \theta_{H(chem)}]^2. \tag{6.39}$$

This means that an exponential decrease of the hydrogen absorption rate occurs and v/v^0 approaches finally zero. The exposure time for severe contamination, $\tau = 1/k_{CO+}p_{CO}$, depends on the CO partial pressure and the reaction coefficient k_{CO+} which is normally close to the maximum reaction rate or sticking coefficient, r_{max}, in the range of 10^9 ML bar^{-1}s^{-1}.

Poisoning by blocking of reactive surface sites can be easily incorporated also in the advanced model for hydriding, described in Sect. 4.4, if the relations (6.36–6.39) are inserted in the rate equations of surface reactions. The use of a model with only one rate determining step is a reliable approximation for poisoning since the reactions normally become dissociation controlled if the contamination of the surface is progressing.

6.4.2 Poisoning by Oxide Layers

The experimental data presented in Sect. 6.1 demonstrate that the surface blocking model does not describe adequately the poisoning phenomena caused by oxygen since severe poisoning is observed in many systems only for coatings thicker than one monolayer equivalent. At least two parameters are needed for the simulation of the experimental findings. One parameter has to define the region of soft poisoning which is observed before oxidation curves have reached the end of the plateau length (Fig.5.7) and the other must describe the steep reduction of the hydrogen absorption rate when the stage 1 in the oxidation models is exceeded and the oxidation rate is retarded drastically. In this section it will be discussed how this can be managed by simple models with reasonable assumptions for the reaction mechanism.

Oxide layer growth

The models proposed here contain the following restricting assumptions. The oxide layers shall be dense and no cracking or spalling of the coating shall occur. Furthermore, the dissolution of the oxide scale in the metal at elevated temperature, discussed in Sect. 6.2, is not considered. The oxygen absorption rate shall be first order with respect to pressure and only a function of the amount of absorbed oxygen per surface area, N_O. It shall not depend on the amount of absorbed hydrogen, N_H, or on time. Under these conditions the absorption rate of oxygen can be described as

$$\frac{dN_O}{dt} = k(T) \cdot r(N_O) \cdot p_{O_2}. \tag{6.40}$$

with $r(N_O)$ the reaction probability. The curves shown in Fig. 5.7 can be simulated by functions $r(N_O)$ which have in the most simple approximation the form

$$r(N_O) = 1 \qquad \text{for} \quad N_O < N_{O,\text{plateau}}, \quad \text{and}$$
$$r(N_O) = \exp(-k\, N_O) \qquad \text{for} \quad N_O > N_{O,\text{plateau}}. \tag{6.41}$$

One can also use other, more smooth functions, for example the function given by (6.1).

In a numeric treatment the amount $N_O(t)$ of absorbed oxygen after an exposure time t, or in other words, the layer thickness represented in units of monolayer equivalents, is given by the relation

$$\int_0^{N_O} \frac{dN_O}{r(N_O)} = k \int_0^t p_{O_2}(t)\, dt \tag{6.42}$$

or for constant O_2 exposure pressure

$$\int_0^{N_O} \frac{dN_O}{r(N_O)} = k p_{O_2} \cdot t. \tag{6.43}$$

Hydrogen absorption rate

From the general trends observed by experiments it can be concluded that poisoning of the hydrogen reaction by oxide scales is mainly a problem of impeded dissociation of physisorbed hydrogen molecules on the oxide surface. Furthermore, results from the experiments shown in Sect. 6.1 and from theory [6.24] indicate that dissociation is enhanced if free or weakly bonded electrons are present on the sample surface. Thus, one can formulate the partial steps responsible for dissociation by the following reaction scheme:

$$H_2(\text{phys}) + 2\,\text{sites} + \text{electron} \rightarrow H_2^-(\text{activated complex}),$$
$$H\text{-}(\text{activated complex}) \rightarrow 2H(\text{chem}) + \text{electron}, \tag{6.44}$$

or

$$H_2(\text{phys}) + 2\,\text{sites} + O^-(\text{chem}) \rightarrow H_2^-(\text{activated complex}) + O(\text{chem}),$$
$$H_2^-(\text{activated complex}) + O(\text{chem}) \rightarrow 2H(\text{chem}) + O^-(\text{chem}).$$

$$(6.45)$$

If electrons from the metal phase or from a metallic hydride phase are available the dissociation process on the sample surface does not depend on the amount of hydrogen absorbed. This hypothesis is supported by the observation that the reaction probability for hydrogen on clean metal surfaces is close to one at low pressure. On hydrides, where less electrons are available, the reaction rate is reduced somewhat but it still depends not on the amount of absorbed hydrogen or, in other words, on the hydride layer thickness (Figs. 6.1, 4). The catalytic substances nickel, iron, and palladium form no hydrides in a hydrogen atmosphere.

The situation on an oxide surface is different. In stage 1 of the oxidation models discussed in Chap. 5 surface charges are present which are responsible for the creation of the Mott equilibrium potential. As long as this potential V_{Mott} is constant the concentration of the surface charges decreases proportionally to the thickness of the oxide layer (B.4). Thus, the hydrogen dissociation rate should obey a relation

$$v \propto v^0 \cdot V_{\text{Mott}}/l \qquad (l > 1\text{ML}). \qquad (6.46)$$

That means the reaction rate v^0 observed for hydrogen dissociation on an oxygen layer is decreased proportionally to $1/l$ with l the layer thickness in ML. In stage 2 of the oxidation model, the Mott equilibrium potential is decreased exponentially with the oxide layer thickness. Then the concentration of negative surface charges is decreased strongly and the region of severe poisoning is reached. In the volumetric experiments the Mott equilibrium potential is at first low since it has to adapt the reaction rate to the impingement rate of hydrogen molecules. This would be an argument to correlate the availability of electrons on the surface better with the tunneling current and yields an exponential decrease of the dissociation probability.

The qualitative behavior of the hydrogen absorption kinetics on oxidizing surfaces, namely slow decrease of the hydrogen absorption rate in stage 1 of the oxidation process and severe poisoning in stage 2, can be simulated in a model by functions which describe the trend shown in Fig. 6.4. A simple function would be:

$$v_H = v^0/(1 + k_1 N_O) \qquad \text{for} \quad N_O < N_{O,\text{plateau}}$$

and

$$v_H = \frac{v^0}{(1 + k_1 N_{O,\text{plateau}})} \cdot \exp(-k_2(N_O - N_{O,\text{plateau}})) \quad \text{for } N_O > N_{O,\text{plateau}}. \qquad (6.47)$$

The term $1/(1 + k_1 N_O)$ instead of a more precise term $1/k_1 N_O$ has to be chosen to avoid problems for $N_O < 1$ where the model is not exactly defined. Instead of (6.46) or (6.57) also other functions can be chosen with slow decrease of the reaction rate for $N_O < N_{O,\text{plateau}}$ and strong decrease for $N_O > N_{O,\text{plateau}}$. A very simple approach would be the definition

$$v = v^0(1 - k N_O) \qquad \text{for } N_O < N_{O,\text{plateau}}$$

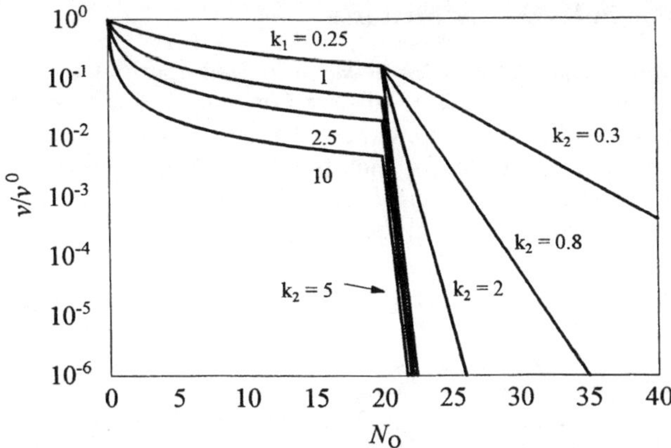

Fig **6.25** Simulation of poisoning curves by the fit function (6.47)

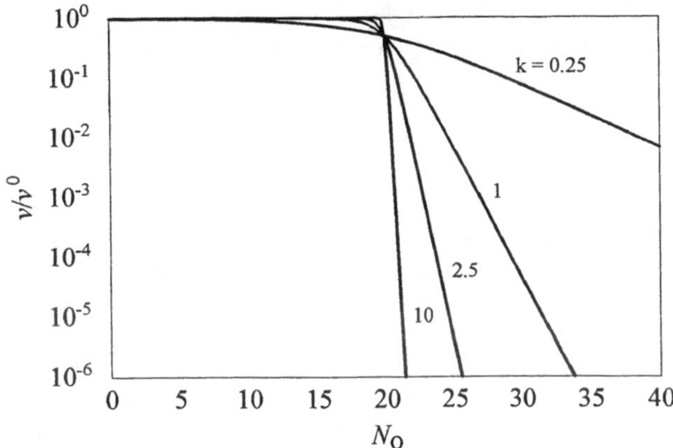

Fig **6.26** Simulation of poisoning curves by the fit function (6.49)

and

$$v = 0 \quad \text{for } N_O \geq N_{O,\text{plateau}'}. \tag{6.48}$$

Another function that can be used is the Fermi function

$$v = v^0[1 + \exp(k\, N_O)]. \tag{6.49}$$

Here, the shape of the curve is given again by two parameters v^0 and k.

Since the exact shape of poisoning curves is normally not determined very precisely, the choice of a special function should not affect much the quality of the fit procedure. This is demonstrated by the set of different fit curves shown in Figs. 6.25, 26.

6.4.3 Hydrogen Absorption in H$_2$/O$_2$ Gas Mixtures

Simulation of exposure experiments in mixtures of H$_2$ with O$_2$ impurities by numeric methods have to determine the amount of absorbed oxygen, ΔN_O, for each time interval from the beginning of an exposure experiment and to add this to the total amount of absorbed oxygen, N_O. Then, for each time interval the absorption rate $v(N_O)$ has to be determined as a function of N_O. The amount of absorbed hydrogen as a function of time is obtained from the relation

$$\frac{dN_H/N_m}{dt} = \frac{F}{V} v_+^0 \cdot f(N_O) \cdot p_{H_2}(1 - \theta_{H(chem)})^2 \tag{6.50}$$

with F/V ratio sample surface to volume, v_+^0 a constant of the system, p_{H_2} the exposure pressure, $(1 - \theta_{H(chem)})$ empty sites for hydrogen chemisorption, and $f(N_O)$ the reaction probability for H$_2$ molecules depending on the amount of poisoning oxygen on the surface.

Equation (6.50) can also be interpreted as the rate law for chemisorption on the oxide surface according to the conditions (6.44) or (6.45). Then $f(N_O)$ is equivalent to a *concentration term* for electrons on the surface. If we stay with this definition and equilibrium is assumed between the hydride subsurface and the hydrogen coverage on the oxide surface, the backward reaction rate can be written as

$$\frac{dN_H/N_m}{dt} = -\frac{F}{V} v_-^0 \, f(N_O) \, k\theta_{H(subsurf.)}^2. \tag{6.51}$$

A simplified poisoning model can be defined when in the advanced reaction model discussed in Sect. 4.4 the partial reaction rates for hydrogen dissociation and recombination on the hydride surface are replaced by (6.50, 51). One has only to include the parallel running contamination reaction into the numerical procedure and insert the $N_O(t)$ values into the equations for the dissociation partial step. As mentioned above, probably no additional information can be gained from advanced models with more rate determining partial steps since poisoned samples have surface controlled kinetics anyway.

6.4.4 Estimate of Exposure Time and H$_2$ Absorption Before Poisoning

For rough estimates the effects of contamination shown, for example, in Fig. 6.4, can be simplified very much. If we assume a reaction probability of one for oxygen and define the plateau length $N_{O,plateau}$, then the amount of absorbed oxygen on the surface is given by

$$N_O = k_{O_2}p_{O_2} \cdot r_{O_2} \cdot t \approx kp_{O_2}t \tag{6.52}$$

and the *contamination time* t_{cont}

$$t_{cont} = N_{O,plateau}/(k_{O_2}r_{O_2}p_{O_2}). \tag{6.53}$$

The hydrogen absorption rate in the period of soft poisoning shall have a mean reaction probability of r_{H_2}'. Then the amount of absorbed hydrogen in the bulk before severe poisoning occurs is given as

$$\frac{N_{\mathrm{H}_2}}{N_{\max}} = \frac{F}{V} k_{\mathrm{H}_2} \cdot p_{\mathrm{H}_2} \cdot r'_{\mathrm{H}_2} \cdot t_{\mathrm{cont}} \tag{6.54}$$

or

$$\frac{N_{\mathrm{H}_2}}{N_{\max}} = \frac{F}{V} \frac{k_{\mathrm{H}_2}}{k_{\mathrm{O}_2}} \cdot \frac{p_{\mathrm{H}_2}}{p_{\mathrm{O}_2}} \cdot \frac{r'_{\mathrm{H}_2}}{r_{\mathrm{O}_2}} \cdot N_{\mathrm{O,plateau}}. \tag{6.55}$$

Equation (6.55) shows that much hydrogen is absorbed before severe poisoning occurs if the specific surface of the sample is large and the oxygen partial pressure is small compared to the hydrogen pressure. The ratio $r'_{\mathrm{H}_2}/r_{\mathrm{O}_2}$ is normally in the range 10^{-2} to 10^{-4}. If the layer thickness of the oxide scale is large ($N_{\mathrm{O,plateau}} \gg 1$) then, of course, also the amount of hydrogen absorbed before poisoning is large.

The amount N_{H_2} obtained by the estimate can be much larger than the amount of hydrogen needed to form the hydride phase, $N_{\max}$. If this number is, for example, 100 then 100 loading unloading cycles can be performed with the sample under the conditions considered before poisoning occurs.

APPENDICES

A. Chemical Potentials and Standard States

A.1 Chemical Potentials

The equilibrium constant of a chemical reaction is given by the sum of the chemical or electrochemical potentials of the species involved (Sect. 2.1). The chemical potential of a species i in a phase is defined as [A.1–10]

$$\mu_i = \mu_i^0(T) + kT \ln a_i. \tag{A.1}$$

The standard potential μ_i^0 is defined by $a_i = 1$ and depends on temperature. The activity a_i describes the concentration and/or pressure dependence of the chemical potential and is an ideal or ideal diluted systems proportional to the mole fraction x_i. In real systems it can depend also on temperature and frequently; it is a function of the concentrations of all species present in the phase.

The units used for the activity define the standard state of a species. They can be chosen according to the convenience of the user. The activity of a species i in a diluted solution is defined as $a_i = \gamma \cdot c_i$ where the concentration c_i can be measured in terms of mol fractions, x_i, of atomic percent, of weight percent, or of any other unit. In this book c_i denotes density of species per m^3. The activity coefficient $\gamma_i(T, c_i, c_j \ldots)$ can be set one if the concentration in the unit chosen is one or for any other concentration value convenient as a definition. Whatever the values of experimental parameters are at a real or virtual state, the state where $a_i = 1$ holds is defined as the standard state of the component i in the phase α. For quantitative thermochemical calculations the standard states of all species in all phases must be exactly defined. Very large errors arise if incompatible units are mixed. The standard states given here are the same as used for the characterization of the species in the models treated in preceding chapters.

A.1.1 Definitions

The expressions for chemical potentials of individual species in different phases are memorized better if some definitions frequently used in thermochemistry are reminded. The *components* of a system are represented by the smallest set of particles which can describe the phases and species of a system. These are normally the chemical elements. If ions are present also electrons have to be introduced as a separate component. All other species occurring in a system such as molecules, ions, lattice defects, or compounds can be defined as a reaction product formed by a reaction of the components acting as reactants. The chemical standard potential of composed species is the Gibb's free reaction enthalpy ΔG^0 of formation if the components are in their standard states.

Most compilations on thermochemical data are based on the definition that the μ value of chemical elements at 298 K and one bar pressure are zero. Since the ΔG^0 values of chemical reactions are given by sums and differences of μ^0 values, this convention is very convenient and provides unambiguous values for the standard Gibbs free reaction enthalpies of compounds. In thermochemistry the following definitions are normally used for the presentation of the standard states for μ^0:

Solid elements:

$$\mu^0(T) = \mu(T) = g(T) = h(T) - Ts(T) \tag{A.2}$$

which is the molal Gibbs free enthalpy.

Solutions:

$$\mu_i = \mu^0(T) + kT \ln(\gamma x_i). \tag{A.3}$$

In this equation deviations from the ideal behavior ($\gamma = 1$) are described by the activity coefficient γ. For the solvent γ becomes unity in the limit $x_i \rightarrow 1$ and Raoult's law is obeyed: $a_i = x_i$.

Ideal dilute solutions:

$$\mu_i = \mu^0 + kT \ln \gamma + kT \ln x_i = \mu^{0'} + kT \ln x_i. \tag{A.4}$$

Here the activity coefficient term is added to the μ^0 term and the standard potential $\mu^{0'}$ is now a quantity which describes the change of the g value of the species i when it is transferred from the standard state to the diluted state in the solvent for the limit $x_i \rightarrow 0$. This virtual state is, of course, not a very realistic situation, even for a dilute solution. Nevertheless, it is a very convenient definition since the simple laws of ideal solutions can be maintained and γ is now a constant depending only on temperature. Especially lattice defects, which are normally present in concentrations well below one percent, are correctly described as ideal diluted components.

Compounds:
For a compound $AB \leftrightharpoons A + B$ the chemical potential is defined as

$$\mu_{AB} = \mu_A + \mu_B. \tag{A.5}$$

Since the molal Gibbs free enthalpy of formation for a stoichiometric compound is almost constant at constant temperature one can write

$$\mu_{AB} \approx g_{AB} = \mu_A + \mu_B. \tag{A.6}$$

In equilibrium the chemical potentials of all components must have the same values in all phases. If, for instance, AB is assumed to be a stable oxide MO, then this phase can be in equilibrium with one bar O_2 gas where $1/2\mu_{O_2}$ is zero by definition. Under this condition

$$\mu_{MO} = \mu_M = g_{MO} \tag{A.7}$$

holds, which is a relatively large negative quantity. On the other hand, in a closed receiver containing a system with metal excess the μ_M value becomes zero and the oxygen partial pressure in the system attains the low value of the oxide decomposition pressure. Then

$$\mu_{MO} = 1/2\mu_{O_2} = g_{MO} \tag{A.8}$$

holds. This example demonstrates that one has to distinguish between the chemical potential μ_{AB} of a stoichiometric compound, which can be used as a concentration independent quantity in the equilibrium constant of a reaction, and the chemical potentials μ_A or μ_B of the components or of defects inside the compound. These values are strongly concentration dependent. However, they are not independent from each other since $\mu_A + \mu_B = \mu_{AB} = g_{AB}$ is constant.

Gas species:
The chemical potential of a species i in the gas phase depends on the partial pressure according to the equation

$$\mu_i = \mu_i^0(T, p = 1) + kT \ln p_i. \tag{A.9}$$

If the total gas pressure is not much higher than one bar the equation for ideal gases is a reliable approximation for gases.

A.2 Standard States and Standard Reactions

This section compiles thermochemical quantities of the species which occur as components or reaction products in the models of the preceding chapters.

A.2.1 Elements

Chemical elements, condensed at 298 K:
For example a metal atom M

$$\mu_M = \mu_M^0(T) + kT \ln a_M, \tag{A.10}$$

$$a_M = 1 \quad \text{for pure metals,}$$

$$\mu_M^0(298) = 0.$$

Chemical elements, in gas molecules at 298 K:
For example O in O_2 molecules

$$\mu_O = \mu_O^0(T) + kT \ln a_O, \tag{A.11}$$

$$a_O = \sqrt{p_{O_2}},$$

$$a_O = 1 \quad \text{for } p_{O_2} = 1 \text{ bar},$$

$$\mu_O^0(298) = 1/2\mu_{O_2}^0(298) = 0.$$

Electrons:

$$\eta_e = \mu_e - e\varphi = \mu_e^0 + kT \ln a_e - e\varphi, \tag{A.12}$$

$$a_e = 1 \quad \text{for electrons at the Fermilevel},$$

$$\mu_e^0 = E_{\text{Fermi}} \text{ (in the metal)}.$$

The standard state of electrons where $a_e = 1$ holds is normally defined as the position in infinite distance from the surface of the condensed phases under consideration. Another standard state which is more convenient for the discussion of metal-oxygen systems and used in this book is the Fermi level of the metal phase [A.10]. The variable $\varphi = \varphi_{\text{Galvani}}$ is the potential at a position inside a phase, on an interphase, or in the gas phase with respect to the reference state at the Fermilevel in the metal.

A.2.2 Oxides M_xO_y

Standard reaction:

$$xM + y/2\,O_2 = M_xO_y, \tag{A.13}$$

$$\mu_{M_xO_y} = \mu_{M_xO_y}^0(T) + kT \ln a_{M_xO_y},$$

$$a_{M_xO_y} = a_M^x \cdot a_O^y \cdot \exp(-\Delta G_{M_xO_y}^0/kT),$$

$$\Delta G_{M_xO_y}^0 = \mu_{M_xO_y}^0 - x\mu_M^0 - y\mu_O^0,$$

$$\Delta G^0 = \Delta H^0 - T\Delta S^0.$$

ΔH^0 is an almost constant and relative large negative quantity in the range -0.3 to -6 eV per oxygen atom and the standard reaction entropy ΔS^0 is about 10^{-3} eV/K.

Activities:

The activity of the oxide, $a_{M_xO_y}$, is normally set to unity. This has the consequence that the product

$$a_M^x \cdot a_O^y = \exp(\Delta G^0_{M_xO_y}/kT) = K(T) \tag{A.14}$$

must be constant with the limiting cases:

Equilibrium with the gas phase

$$a_O = \sqrt{p_{O_2}}, \tag{A.15}$$

$$\ln a_M = \Delta G^0/xkT - y/2x \cdot \ln p_{O_2}.$$

Equilibrium with the metal

$$a_M = 1, \tag{A.16}$$

$$\ln a_O = \Delta G^0/ykT,$$

or

$$1/2 \ln p^*_{O_2} = \Delta G^0/ykT.$$

The quantity $\sqrt{p^*_{O_2}}$ is the activity of oxygen in the oxide; p^* is measured in bar. It is for most oxides a very small number and far below a realistic pressure accessible by experiment for decomposition.

The equations and definitions for other metal-gas compounds such as hydrides, nitrides, or chlorides are the same. If equilibria between two or more oxides are considered, the oxygen activity is the same in all phases and defines the O_2 equilibrium pressure of the system. The metal activities in the oxides can be obtained from (A.14).

A.2.3 Physisorption

Standard reaction:

$$O_2 \leftrightarrows O_2(\text{phys}),$$

$$\mu_{O_2(\text{phys})} = \mu^0_{O_2(\text{phys})} + kT \ln a_{O_2(\text{phys})}. \tag{A.17}$$

The quantity $\Delta G^0 = \Delta H^0 - T\Delta S^0$ of ideal physisorption systems depends only on temperature. The coverage θ_{O_2} is defined as the ratio of the occupied physisorption sites, $N_{O_2(\text{phys})}$, to the total number of sites, N,

$$\theta_{O_2} = N_{O_2(\text{phys})}/N, \tag{A.18}$$

$$a_{O_2(\text{phys})} = \theta_{O_2}/(1 - \theta_{O_2}).$$

The equilibrium constant then becomes

$$\frac{\theta_{O_2}}{(1 - \theta_{O_2}) \cdot p_{O_2}} = \exp(-\Delta G^0/kT),$$ (A.19)

$$\Delta G^0 = \mu^0_{O_2(\text{phys})} - \mu^0_{O_2(\text{gas})},$$

$$\theta_{O_2(\text{phys})} = \frac{p_{O_2} \cdot \exp(-\Delta G^0/kT)}{1 + p_{O_2} \cdot \exp(-\Delta G^0/kT)}.$$ (A.20)

Equation (A.20) shows that for $p_{O_2} \cdot \exp(-\Delta G^0/kT) \ll 1$

$$\theta_{O_2(\text{phys})} \propto p_{O_2}$$ (A.21)

holds and for $p_{O_2} \to \infty$

$$\theta_{O_2(\text{phys})} = 1.$$ (A.22)

The physisorption enthalpy ΔH^0 is a small negative quantity of about -0.1 to -0.5 eV and ΔS^0 is about $1.5 \cdot 10^{-3}$ eV/K. Other gases such as H_2, N_2 or H_2O have comparable ΔH^0 and ΔS^0 values.

A.2.4 Molecular Chemisorption

Standard reaction:

$$O_2(\text{gas}) + ze \leftrightarrows O_2^{z-},$$

$$\eta_{O_2^{z-}} = \mu^0_{O_2^{z-}} + kT \ln a_{O_2^{z-}} - ze\varphi_{\text{surface}}.$$ (A.22)

The variable φ_{surface} is the potential difference between the surface site of the chemisorption species and the position of electrons in the standard state. In the oxidation models this is the Fermilevel of the metal phase.

For ideal systems with constant ΔG^0, ΔH^0, and ΔS^0 values for only one type of chemisorption sites the following equations are used:

$$a_{O_2^{z-}} = \theta_{O_2^{z-}}/(1 - \theta_{O_2^{z-}}),$$

$$\frac{\theta_{O_2^{z-}}}{(1 - \theta_{O_2^{z-}})p_{O_2}} = \exp(-\Delta G^0/kT) \cdot \exp(ze\varphi_{(\text{surface})}/kT),$$ (A.23)

$$\Delta G^0 = \mu^0_{O_2^{z-}} - \mu^0_{O_2(\text{gas})} - z\mu^0_e.$$

A.2.5 Atomic Chemisorption

Standard reaction:

$$1/2 O_2(\text{gas}) + ze \leftrightarrows O^{z-},$$

$$\eta_{O^{z-}} = \mu^0_{O^{z-}} + kT \ln a_{O^{z-}} - ze\varphi_{\text{surface}}. \tag{A.24}$$

For ideal systems we obtain

$$a_{O^{z-}} = \theta_{O^{z-}}/(1 - \theta_{O^{z-}}),$$

$$\frac{\theta_{O^{z-}}}{(1 - \theta_{O^{z-}}) \cdot \sqrt{p_{O_2}}} = \exp(-\Delta G^0/kT) \cdot \exp(ze\varphi_{\text{surface}}/kT), \tag{A.25}$$

$$\Delta G^0 = \mu^0_{O^{z-}} - 1/2\mu^0_{O_2(\text{gas})} - z\mu^0_e.$$

z can be zero, one or two. The chemisorption enthalpies ΔH^0 have much larger negative values than the physisorption enthalpies.

A.2.6 Metal Interstitials

Standard reaction:

$$M \leftrightarrows M^{z+} + ze, \tag{A.26}$$

$$\eta_{M^{z+}} = \mu^0_{M^{z+}} + kT \ln a_{M^{z+}} + ze\varphi_{\text{oxide}}.$$

The variable φ_{oxide} is the potential difference between the site of the intersitial in the oxide and the potential of the electron in the standard state. Since the concentration of lattice defects in a compound is a small quantity, the ideal dilute solution is a very convenient standard state. For the treatment of diffusion problems defect concentrations must be given as species per volume, for example, as defects/m^3. In the models presented in preceding chapters, concentrations are given in units $\theta_{M^{z+}}$ of defects per lattice sites in lattice planes perpendicular to the diffusion flux. This quantity provides a more clear idea of relative concentrations and fluxes.

$$a_{M^{z+}} = \theta_{M^{z+}},$$

$$\frac{\theta_{M^{z+}}}{a_M} = \exp(-\Delta G^0/kT) \cdot \exp(-ze\varphi_{\text{oxide}}/kT), \tag{A.27}$$

$$\Delta G^0 = \mu^0_{M^{z+}} - \mu^0_M + z\mu^0_e.$$

$a_M = 1$ holds for pure metals at the interface oxide/metal and

$a_M = \exp(\Delta G^0_{M_x O_y}/xkT) = \left(\sqrt{p^*_{O_2}}\right)^{y/x}$ at the oxide surface if p_{O_2} is about one bar (A.15, A.16).

A.2.7 Metal Vacancy

Standard reaction:

$$ze \leftrightarrows M + [M]^{z-},\tag{A.28}$$

$$\eta_{[M]^{z-}} = \eta^0_{[M]^{z-}} + kT \ln a_{[M]^{z-}} - ze\varphi_{\text{oxide}}.$$

The potential φ_{oxide} and the concentration terms are defined in Sect. A.2.6. For the ideal diluted system we get

$$a_{[M]^{z-}} = \theta_{[M]^{z-}},$$

$$\theta_{[M]^{z-}} \cdot a_M = \exp(-\Delta G^0/kT) \cdot \exp(ze\varphi_{\text{oxide}}/kT),\tag{A.29}$$

$$\Delta G^0 = \mu^0_{[M]^{z-}} + \mu^0_M - z\mu^0_e.$$

Again, $a_M = 1$ holds at the interface oxide/metal and $a_M = \left(\sqrt{p^*_{O_2}}\right)^{y/x}$ at the oxide surface for $p_{O_2} = 1$ bar.

A.2.8 Oxygen Interstitials

Standard reaction:

$$1/2 O_2 + ze \leftrightarrows O^{z-},\tag{A.30}$$

$$\eta_{O^{z-}} = \mu^0_{O^{z-}} + kT \ln a_{O^{z-}} - ze\varphi_{\text{oxide}}.$$

The potential φ_{oxide} and the concentration term are defined in Sect. A.2.6. For the ideal diluted system one can write

$$a_{O^{z-}} = \theta_{O^{z-}},$$

$$\frac{\theta_{O^{z-}}}{\sqrt{a_{O_2}}} = \exp(-\Delta G^0/kT) \cdot \exp(-ze\varphi_{\text{oxid}}/kT),\tag{A.31}$$

$$\Delta G^0 = \mu^0_{O^{z-}} - 1/2\mu^0_{O_2} - z\mu^0_e.$$

At the interface metal/oxide $a_O = \sqrt{p^*_{O_2}}$ holds and at the oxide surface $a_O = \sqrt{p_{O_2}}$.

A.2.9 Oxygen Vacancy

Standard reaction:

$$0 \leftrightarrows 1/2\,O_2 + [O]^{z+} + ze, \tag{A.32}$$

$$\eta_{[O]^{z+}} = \mu^0_{[O]^{z+}} + kT \ln a_{[O]^{z+}} + ze\varphi_{\text{oxide}}.$$

φ_{oxide} and the concentration are defined in section A.2.6. For the ideal diluted system one obtains

$$a_{[O]^{z+}} = \theta_{[O]^{z+}},$$

$$\theta_{[O]^{z+}} \cdot \sqrt{p_{O_2}} = \exp(-\Delta G^0/kT)\exp(-ze\varphi_{\text{oxide}}/kT), \tag{A.33}$$

$$\Delta G^0 = \mu^0_{[O]^{z+}} + 1/2\mu^0_{O_2} + z\mu^0_{e}.$$

At the interface metal/oxide $a_O = \sqrt{p^*_{O_2}}$ holds and at the oxide surface $a_O = \sqrt{p_{O_2}}$.

A.2.10 Other Compounds

Other compounds than oxides are treated in the same way. In the metal-like phases of hydrides, carbides or nitrides no charged defects are present and, thus, z is zero. The electrochemical potentials η are then identical with the chemical potentials μ.

B Equilibria of Charged Species

Oxide layers are normally semiconducting phases composed of ionic components. The growth rate is determined by the flux of charged lattice defects, which can be enhanced or impeded strongly by electric fields. The fields existing in an oxidation system under steady state conditions are produced by charges at interfaces and in the bulk of the oxide. Due to the high electronegativity of oxygen atoms, excess charges present at the oxide surface are normally negative. In dry oxidation they must be compensated by positive charges in the oxide layer or at the metal/ oxide interface. In contrast to electrochemical reactions the electric field is not an experimental parameter which can be controlled from outside, but it is established by reactions between charged particles under the restricting conditions of charge neutrality and coupled currents. For each of the partial reactions equilibrium conditions exist which are attained in the limit of zero reaction rate. They can, of course, only be established if the forward and backward rates of the partial step considered are much faster than the rates of competing reactions in the system.

In this section some aspects of equilibria between charged species are discussed which may improve the understanding of the models presented in previous chapters. Emphasis is laid especially on the problem of space charges in thin semiconducting layers which are small when compared with the screening length. A screening length larger than the oxide layer thickness is one of the preconditions necessary for fast Mott-potential enhanced oxidation at low temperature.

B.1 Poisson Equation

The correlation between electric charges and the change of the electric field is given by the Poisson equation

$$\Delta \varphi = -\frac{\rho}{\epsilon \cdot \epsilon_0}.$$ (B.1)

With space charge density $\rho = z_i e c_i$, concentration c_i in units charged particles/ m^3, ϵ_0 permittivity of vacuum and ϵ relative permittivity. For a plane system where charged particles are distributed homogeneously on an infinite lattice plane of thickness a with a relative density θ_i of charged particles in ML one can write

$$-\frac{d^2\varphi}{dx^2} = \frac{dE}{dx} = \frac{\Delta E}{a} = \frac{z_i e \cdot \theta_i}{\epsilon \cdot \epsilon_0} \approx 1.8 \cdot 10^{11} \cdot \frac{z_i \theta_i [\text{ML}]}{\epsilon} \, [V/m].$$ (B.2)

The maximum areal density of charged particles is one ML or about 10^{19} particles/m^2. This relation shows that in a dielectric material at each lattice plane with electric excess charges the field is changed by an amount of ΔE.

B.2 Dipole Layers

Between two planes occupied by charges of opposite sign a constant field E and a potential difference $\Delta\varphi$ exist,

$$|E| = \frac{z_i e \cdot \theta_i}{\epsilon \cdot \epsilon_0}, \tag{B.3}$$

$$|\Delta\varphi| = \frac{z_i e \cdot \theta_i}{\epsilon \cdot \epsilon_0} \cdot l \approx 10^{11}\,\frac{z_i \theta_i [\text{ML}]}{\epsilon} \cdot l\,[\text{m}] \cdot [\text{V/m}]. \tag{B.4}$$

For a dipole layer with $z_i\theta_i/\epsilon \approx 0.1$ to 1 and $l \approx 3 \cdot 10^{-10}$m the potential difference $\Delta\varphi$ is about 1 to 10 V and comparable to energies of chemical bonds or potential differences in electrochemical cells. With increasing thickness l the surface charge θ decreases proportionally to $1/l$ if the potential difference $\Delta\varphi$ is kept constant.

B.3 Space Charges

In this section the problem of space charges in dielectric films of limited thickness is considered in more detail because this is a crucial point in low temperature oxidation. Equations for the electric field, space charge concentrations and potentials will be derived for a very simple but informative model. We look at a system where inside a semiconducting layer charged particles are formed by a reaction at the metal/oxide interface and charges of opposite sign are present on the other side of the phase boundary in the metal. Such a situation can exist, for example, if positive oxygen vacancies at the metal oxide interface are formed,

$$\text{M} \leftrightarrows \text{MO} + [\text{O}]^{z+} + \text{e}\,(\text{in M}). \tag{B.5}$$

It is assumed that no surface reactions with the gas phase occur and that no other charged particles, for example excess electrons, are present in the oxide. This treatment is analogous to considerations on electrode reactions in electrochemical cells.

The electrochemical potential of a particle with positive charge is written as

$$\eta = \eta^0(T) + kT \ln c + ze\varphi. \tag{B.6}$$

In equilibrium the electrochemical potential inside a phase is constant. If the concentration c, given here in particles/m^3, and the potential φ depend only on local coordinates then the concentration and the electric field are interrelated by the equation

$$\nabla \ln c = -\frac{ze}{kT}\nabla\varphi = k_e E \tag{B.7}$$

with $k_e = ze/kT$. Inserting this in the Poisson equation (B.1) yields

$$\Delta \ln c = k_c c = k_c \exp(\ln c) \tag{B.8}$$

with

$$k_c = \frac{e^2}{\epsilon_0 k} \cdot \frac{z^2}{\epsilon T}.$$

The mathematical treatment is simplified enormously if the considerations are restricted to a plane system where c and φ depend only on the x coordinate. Integration of (B.8) is accomplished by setting $\gamma = \ln c$ and $\partial \ln c / \partial x = \partial \gamma / \partial x = p$:

$$\int_{p_0}^{p} p \, dp = k_c \int_{\gamma_0}^{\gamma} \exp(\gamma) \, d\gamma,$$

$$\left(\frac{\partial \ln c}{\partial x} \right)^2 - 2k_c c = \left(\frac{\partial \ln c_0}{\partial x} \right)^2 - 2k_c c_0. \tag{B.9}$$

Equation (B.9) correlates the concentration c and the concentration gradient $\nabla \ln c$ at each position inside a semiconducting phase. The correlation function depends on two parameters, the maximum space charge density c_0 and $\ln c_0$ at one of the phase boundaries. The assumptions made so far are that the electrochemical potential η is constant inside the oxide phase and that Poisson's equation (B.1) holds. Equation (B.9) can be rewritten as

$$\frac{d \ln c}{dx} = -\sqrt{2k_c c_0} \cdot \sqrt{\kappa - 1 + c/c_0} \tag{B.10}$$

with

$$\kappa = \left(\frac{d \ln c_0}{dx} \right)^2 / 2k_c c_0.$$

In our model the space charge concentration c_0 has the maximum value at the metal/oxide interface $x = 0$. The sign of $\partial \ln c / \partial x$ is negative and because the expression under the square root must be positive $0 < \kappa < 1$ holds. ∇c decreases from the maximum value ∇c_0 and becomes zero for $c/c_0 = (\kappa - 1)$. ∇c approaches zero for $c = 0$ only if κ is one. In addition the electric field must disappear at the oxide surface because of the condition of charge neutrality. Since $E = 1/k_c \cdot \nabla \ln c = 0$ holds also $\nabla \ln c$ must be zero. This condition does not depend on the thickness of the layer. $\nabla \ln c$ and ∇c must be zero in any case at the end of the oxide layer $(x = l)$ opposite to the interface where the defects are formed and the quantity κ can be defined as (B.10)

$$\kappa = 1 - c_l/c_0. \tag{B.11}$$

In models on low temperature oxidation we have to distinguish layers which are thicker than the *screening length* λ,

$$\lambda = 1/\sqrt{2k_c c_0} \approx \frac{\sqrt{\epsilon} \cdot 10^{-11}}{z \sqrt{x_i}} \, [\text{m}] \tag{B.12}$$

from those which are thinner. Concentrations are given in (B.12) in addition to c_i also in mol fractions x_i in order to provide a more clear idea of the screening length which extends, roughly spoken, to a length of about $x^{-1/2}$ atomic distances.

B.3.1 Thick Layers, $l \gg \lambda$

This limit for an oxide layer thickness l larger than $\lambda = l/\sqrt{2k_c c_0}$ is approached for $\kappa = 1$ (Fig. B.1a). Integration of (B.10) yields

$$c = c_0/(1 + x/2\lambda)^2. \tag{B.13}$$

The screening length $\lambda = 1/\sqrt{2k_c c_0}$ plays a fundamental role in any system with mobile charged particles. In our example at the distance λ from the metal/oxide interface in Fig. B.1a the concentration c is about the half of the c_0 value. Thus, the screening length λ characterizes the extension of the space charge layer inside a semiconductor. It is proportional to one over the square root of the density of charged particles c_0 at the reactions front.

B.3.2 Thin Layers, $l < \lambda$

If the concentration c_0 of mobile charges is in the range $10^{28}/m^3$, for example electrons in a metal matrix, λ is about 10^{-11}m and the space charge layer can be approximated by an interface charge of atomic dimensions. However, if c_0 is about 10^{24} or 100 at.ppm or even less the screening length λ becomes 10 nm or more and exceeds the thickness of oxide layers formed on metals at room temperature (Fig. B.1 c). In this case the mathematical solutions of the problem have to consider the length l of the oxide layer as an additional parameter.

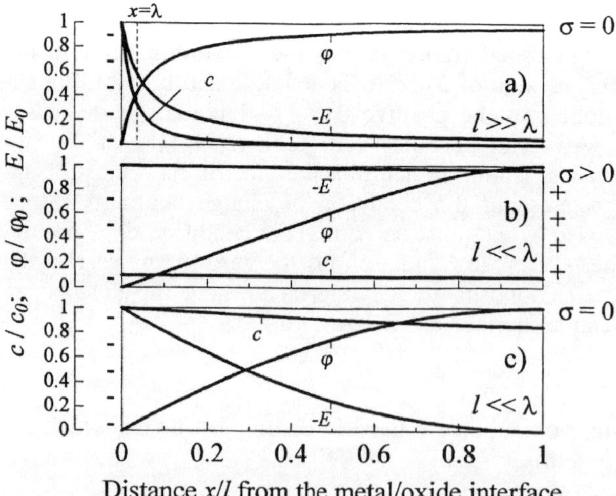

Fig. B.1 Potential φ, electric field E, and space charge density r of oxide layers. a) no surface charges, $l \gg \lambda$; b) charges on both interfaces, $l \ll \lambda$; c) no surface charges, $l \ll \lambda$

The determination of the field E caused by defect formation at an interface is accomplished by equating expressions obtained from the reaction equation (B.5) and (B.10). Equation (B.5) yields the Gibbs free enthalpy of formation ΔG:

$$\Delta G = \Delta G^0 + kT \ln x_i + ze(\varphi_{\text{oxide}} - \varphi_{\text{metal}}). \tag{B.14}$$

ΔG^0 is the standard Gibbs free reaction enthalpy if x_i is taken as mole fraction. In equilibrium $\Delta G = 0$ holds or

$$\varphi_{\text{oxide}} - \varphi_{\text{metal}} = -\frac{\Delta G^0}{ze} - \frac{kT}{ze} \ln x_i. \tag{B.15}$$

If we set $E^0 = -\varphi_1/a$ with a the distance of the charge layer in the metal from the first layer occupied by defects $[O]^{z+}$ in the oxide, E^0 becomes

$$E^0 = \frac{\Delta G^0}{zea} + \frac{kT}{zea} \ln x_{i0}. \tag{B.16}$$

With (B.7,10) we get

$$E^0 = -\frac{\sqrt{2k_c c_o}\sqrt{\kappa}}{k_e} \approx 10^9 z(x_{i0}\,\kappa/\epsilon)^{1/2}\,[\text{V/m}] \tag{B.17}$$

and by equating (B.16) and (B.17)

$$\frac{\Delta G^0}{kT} + \ln x_{i0} + a\sqrt{2k_c f x_{i0}}\sqrt{\kappa} = 0 \tag{B.18}$$

with f conversion factor from density c_0 to mol fraction x_{i0}. This equation determines the value of x_{i0} if ΔG^0 of the defect formation reaction is known. The parameter κ is in the limit $l \gg \lambda$ one. Figure B.2 shows that for x_{i0} values below 10^{-2} the defect concentration at the metal oxide interface is no longer affected by the screening length $\lambda = 1/\sqrt{2k_c c_0}$. It is only a function of the Gibbs free reac-

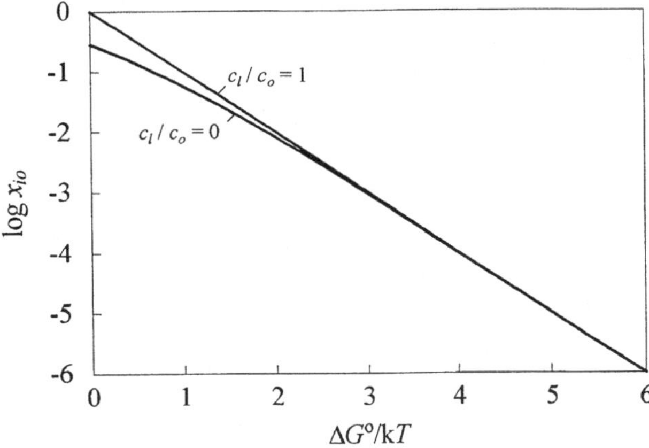

Fig. B.2 Ion concentration x_0 at the metal/oxide interface as a function of the defect formation energy ΔG^0

tion enthalpy ΔG^0. The maximum electric field E^0 for x_0 values below 10^{-4} is smaller than 18^8 [V/m]. If for thin films $l < \lambda$ holds, κ is a small number and the field is reduced still more.

In the limit for thin layers the mathematical treatment of the problem is becoming more complex. The parameter κ, defined by (B.10, 11) has to be treated as an additional variable. Integration of (B.10) for $0 < \kappa < 1$ yields

$$c = c_0(1 - \kappa)[1 + \tan^2(\arcsin\sqrt{\kappa} - \sqrt{k_c c_0(1 - \kappa/2} \cdot x)]. \tag{B.19}$$

In the solutions of interest the correlation between the parameter κ and the thickness l of the oxide layer is defined by the condition that $E = const \cdot d\ln c/dx$ must be zero at the oxide surface $x = l$. One can show that this condition is fulfilled if the following equation holds:

$$l/\lambda = \frac{2\arcsin\sqrt{\kappa}}{\sqrt{1 - \kappa}} = \frac{2\arctan\sqrt{\kappa/(1 - \kappa)}}{\sqrt{1 - \kappa}}. \tag{B.20}$$

This inserted in (B.19) yields

$$c = c_0(1 - \kappa)1 + \tan^2\left[\sqrt{k_c c_0(1 - \kappa)/2} \cdot (l - x)\right]. \tag{B.21}$$

For $x = l$ we get $\kappa = (1 - c_l/c_0)$ and (B.21) becomes

$$\frac{c}{c_0} = \frac{c_l}{c_0}\left[1 + \tan^2\sqrt{\frac{c_l}{c_0}} \cdot \frac{l - x}{2\lambda}\right)\right]. \tag{B.22}$$

Equation (B.20) can then also be written as

$$l/\lambda = \frac{2\arcsin\sqrt{1 - c_l/c_0}}{\sqrt{c_l/c_0}}. \tag{B.23}$$

Figure B.3 shows the concentration profile in a semiconductor in a plot c/c_0 vs. x/l for layers of various values of c_l/c_0 or $1 - \kappa$. These parameters are correlated with the relative thickness l/λ of the layer [(B.23) and Fig. B.4].

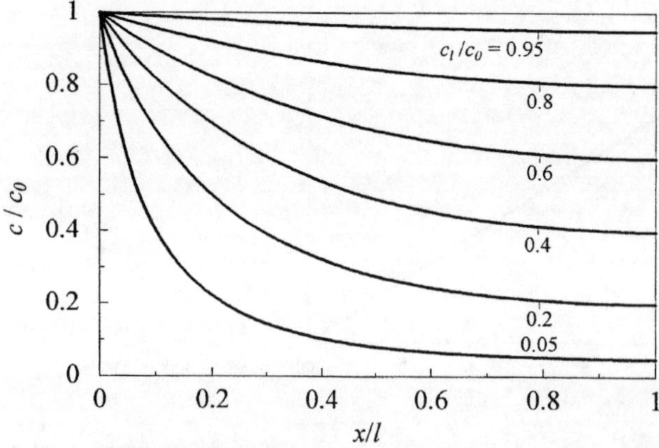

Fig. B.3 Concentration profiles of ions in oxide layers of different relative thickness l/λ

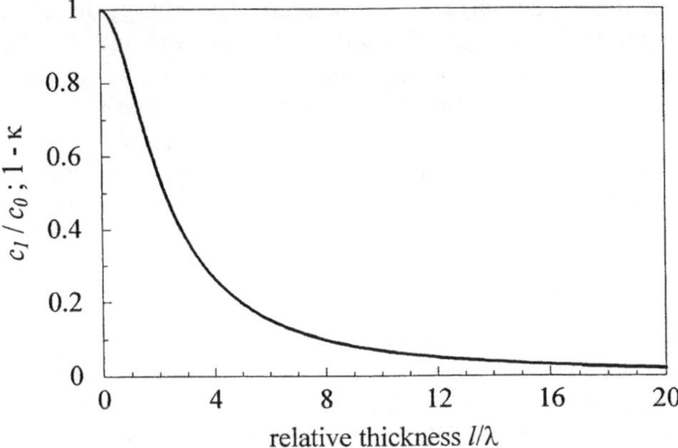

Fig. B.4 Parameter c_l/c_0 or $1 - \kappa$ as a function of the relative oxide layer thickness l/λ

B.3.3 Conclusions

The maximum value of the electric field E^0 at the metal/oxide interface inside the oxide is given by (B.17),

$$\left| E^0_{\max} \right| < \frac{kT}{ze} \cdot \frac{1}{\lambda} = \frac{26}{z\lambda} [\text{mV}] \quad \text{at 300 K.} \tag{B.24}$$

This field, produced by the diffusion potential of space charges, is relatively weak. It is sufficient to prevent further diffusion of defects from the reaction front at the metal/oxide interface to the bulk of the layer, but it cannot remarkably affect the reaction rates in the early stage of low temperature oxidation. There are two reasons for this:

(i) The retardation of a diffusion flux by the field does not go beyond the zero value. Thus, also enhancement cannot exceed a factor of about two and is unable to explain the high initial oxidation rates observed at low temperatures.

(ii) If the oxide layer thickness is smaller than the screening length the field is reduced still further by a factor of about l/λ (Fig. B.3).

The diffusion potentials and space charges in the bulk of the oxide layers are, therefore, in models on low-temperature oxidation much less important effects than in models on semiconductor electronics or on high-temperature oxidation. Commonly they can be disregarded without losing much information on the main features of the mechanism simulated.

B.4 Mott Potential

Oxygen present in the gas atmosphere can be chemisorbed as a molecule or as an atom on the oxide surface. As highly electronegative species, it acts as electron acceptor to form negative ions. If the energy of an electron in such an acceptor site on the surface lies below the Fermi-level of electrons in the metal on the opposite side of the oxide layer, electrons from the metal will migrate to these levels and a negatively charged layer on the oxide surface is produced. The energy level of electrons in the acceptor site, $W = eV_{\text{Mott}}$, is called *Mott potential* and can be correlated with the ΔH^0 value of the standard reaction for the transfer of an electron from the metal phase to the chemisorption species acting as acceptor.

The *equilibrium potential* $\Delta\varphi$ across an oxide layer of the thickness l is a function of the Mott potential W and the concentration of surface acceptor levels for electrons. It is determined by the equilibrium constant of the electron transfer reaction,

electron (metal) + acceptor $\leftrightarrows$ electron (acceptor).

In quantitative discussions one has to distinguish between the Mott potential as an energy level for electrons on the oxide surface, called W in the models of Chap. 5, and the electronic equilibrium potential $\Delta\varphi$ produced by this energy difference across the oxide layer. In most cases the equilibrium potential $\Delta\varphi$ is very similar to the quantity $W/e = V_{\text{Mott}}$. For qualitative considerations it may, therefore, be permitted to use the historically incorrect, but informative term *Mott equilibrium potential* or *Mott potential* also for $\Delta\varphi$ instead of the more exact, but lengthy expression „electric equilibrium potential across the oxide layer caused by acceptor levels on the oxide surface with a distance $eV_{\text{Mott}} = W$ below the Fermilevel in the metal".

A third potential term which has to be distinguished is the „*dynamic potential difference*", V in Chap. 5, established across the oxide layer as a consequence of the coupled current conception which guarantees charge neutrality in the system. If the electronic forward and backward fluxes are much bigger than the ionic fluxes, the dynamic potential V becomes equal to the equilibrium potential $\Delta\varphi$.

The pressure dependence of the electronic equilibrium states is controlled by the kind of acceptor sites on the surface. Three cases have to be distinguished, namely, adsorbed O_2 molecules, or adsorbed oxygen atoms in equilibrium with the gas phase, and acceptor levels on the surface which are not pressure dependent. The latter ones can be attributed, for example, to changes in the valency of metal ions on the oxide surface.

B.4.1 Oxygen Molecules as Acceptors

Reaction equation:

$$O_2(\text{surface}) + ze + \leftrightarrows O_2^{z^-} ,$$

$$kT \ln \frac{\theta_{O_2^{z^-}}}{(1 - \theta_{O_2^{z^-}})\theta_{O_2(\text{phys})}} = -\Delta G_1^0 + ze\Delta\varphi, \tag{B.25}$$

with $\Delta G_1^0 = (\mu_{O_2^{z^-}}^0 - \mu_e^0 - \mu_{O_2(\text{phys})}^0)$, which is a negative quantity, and $z = 1$ or 2. The surface coverage with O_2 molecules, $\theta_{O_2(\text{phys})}$, can be taken from (A.19) or in the limit $\theta_{O_2(\text{phys})} \ll 1$ as $\theta_{O_2(\text{phys})} = p_{O_2}\exp(-\Delta G_2^0/kT)$. Thus, one can write for $\theta_{O_2(\text{phys})} \ll 1$ or for not too high p_{O_2} pressures

$$kT \ln \frac{\theta_{O_2^{z^-}}}{(1 - \theta_{O_2^{z^-}})p_{O_2}} = -\Delta G_3^0 + ze\Delta\varphi \tag{B.26}$$

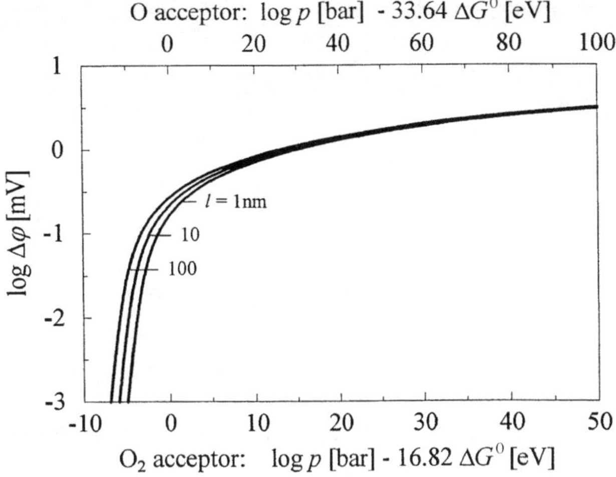

Fig. B.5 Mott equilibrium potential $\Delta\varphi$ as a function of the O_2 pressure and the energy of the electron acceptor level at the surface, ΔG^0

with $\Delta G_3^0 = \Delta G_1^0 + \Delta G_2^0$ and $\Delta\varphi = \varphi_{\text{surface}} - \varphi_{\text{metal}}$ which is the equilibrium potential. The standard potential of electrons μ_e^0 in (B.25) is defined as the Fermi energy E_F in the metal (A.12). The term $\theta_{O_2^{z-}}$ represents a negative surface charge of $z\theta_{O_2^{z-}} \cdot N$ charges on the oxide surface where $\theta_{O_2^{z-}} \cdot N$ is given in ML (1 ML $\approx$ $10^{19}/\text{m}^2$). This charge is balanced by an identical positive layer in the metal at the metal/oxide interface in a distance l. According to (B.4) this is a capacitor charged to a potential difference

$$\Delta\varphi = -l\frac{ze \cdot \theta_{O_2^{z-}}}{\epsilon\epsilon_0} \tag{B.27}$$

or

$$\Delta\varphi \approx -\frac{zl\,[\text{m}]}{\epsilon} \theta_{O_2^{z-}}[\text{ML}] \cdot 10^{11}\,[\text{V}]. \tag{B.28}$$

Equation (B.27) inserted in (B.26) yields

$$\log\left(\frac{|\Delta\varphi|}{l \cdot K_1 - |\Delta\varphi|}\right) + \frac{ze}{2.3 \cdot kT}|\Delta\varphi| - \log p_{O_2} = \frac{-\Delta G_3^0}{2.3 \cdot kT} \tag{B.29}$$

with $K_1 = ze/\epsilon\epsilon_0$. For $z = 1$ the potential $\Delta\varphi$ is plotted in Fig. B.5 as a function of $\log p_{O_2} - 16.82 \cdot \Delta G^0[\text{eV}]$ for different values of the oxide film thickness l at 300 K. It increases with increasing O_2 pressure and for $\Delta\varphi$ values below –0.5 V with layer thickness.

B.4.2 Oxygen Atoms as Acceptors

Reaction equation:

$$O\ (\text{surface}) + ze + \text{site} \leftrightarrows O^{z-},$$

$$kT \ln \frac{\theta_{O^{z-}}}{(1 - \theta_{O^{z-}})\,\theta_{O(\text{chem})}} = -\Delta G_1^0 + ze\Delta\varphi. \tag{B.30}$$

For reversible atomic chemisorption the equilibrium constant is given by (A.25) and in the limit $\theta_O \ll 1$ as

$$\theta_{O(\text{chem})} = \sqrt{p_{O_2}}\,\exp(-\Delta G^0/kT). \tag{B.31}$$

A similar treatment as shown in the previous section yields for the equilibrium potential $\Delta\varphi$ the equation

$$\log\left(\frac{|\Delta\varphi|}{l \cdot K_1 - |\Delta\varphi|}\right) + \frac{ze|\Delta\varphi|}{2.3 \cdot kT} - 1/2 \log p_{O_2} = -\frac{\Delta G_3^0}{2.3 \cdot kT}. \tag{B.32}$$

K_1 can be taken from (B.29). For $z = 1$ curves $\log \Delta\varphi$ vs. ($\log p_{O_2} - 33.64 \cdot \Delta G^0$) are shown also in Fig. B.5 if the upside abscissa scale is used. The curves are then identical with those for O_2^- chemisorption species.

B.4.3 Pressure Independent Acceptor Sites

Reaction equation:

$$A\ (\text{surface}) + ze \leftrightarrows A^{z-},$$

$$kT \frac{\theta_{A^{z-}}}{(1 - \theta_{A^{z-}})} = -\Delta G_1^0 + ze\varphi. \tag{B.33}$$

In this case the equilibrium potential $\Delta\varphi$ becomes

$$\log\left(\frac{|\Delta\varphi|}{l \cdot K_1 - |\Delta\varphi|}\right) + ze\,|\Delta\varphi| = \frac{-\Delta G_1^0}{2.3\,kT}. \tag{B.34}$$

K_1 is given again by (B.29). The $\log \Delta\varphi$ vs. $16.82 \cdot \Delta G_1^0$ curves for this system are the same as those shown in Fig. B.5 if the term $\log p_{O_2}$ at the abscissa is neglected. They depend no longer on p_{O_2} but only on ΔG^0. If the ΔG^0 value is large, the equilibrium potential does not depend strongly on the layer thickness l. This is also demonstrated by the curves $\log \Delta\varphi$ vs. $\log l$ in Fig. B.6. The value of the parameter ($n \log p_{O_2} - \Delta G^0/2.3\,kT$) must be smaller than 10 if the dependence on layer thickness shall become visible. The factor n is $n = 1$ for O_2^- species, $n = 1/2$ for O^- species, and $n = 0$ for electron acceptors which are not associated with adsorbed species in equilibrium with the gas phase.

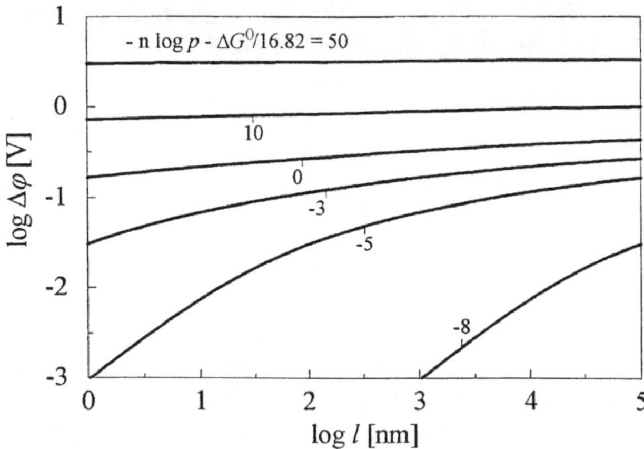

Fig B.6 Mott equilibrium potential $\Delta\varphi$ as a function of the oxide layer thickness l for different values of the energy of the electron acceptor level, ΔG^0

B.4.4 Conditions for the Existence of a Mott Equilibrium Potential

The Mott equilibrium potential is the result of an equilibrium between electrons in the metal phase and acceptor levels on the surface of an oxide scale or another semiconducting coating. It can be established only if the electrons present on the surface of the semiconductor are not consumed by positive species of another fast process proceeding parallel to the electron transfer through the oxide skin. The rate of such reactions is normally strongly enhanced by the Mott field and prevents under normal conditions the establishment of electronic equilibrium. One has to keep in mind that the normal process occurring in a metal/oxygen system in an oxidizing atmosphere at elevated temperatures is the formation of oxides and not merely production of charged oxygen chemisorption species on the oxide surface.

Three requirements are needed for the establishment of a Mott equilibrium potential across an oxide layer:

(i) Fast electronic currents through the layer.

(ii) Acceptor levels for electrons on the oxide surface with energies below the Fermi level in the metal.

(iii) The space charge layer in the semiconducting surface layer must contain in total less charges than the interface layer at the metal/oxide interface.

It is interesting to check which conditions have to be fulfilled by the electric properties of a coating material to meet these requirements. A very important parameter of the problem is the ratio shielding length to layer thickness, λ/l. The shielding length depends on the concentration of mobile charged species in the layer. It is given approximately by the relation (B.12)

$$\lambda \approx 10^{-11} \cdot \frac{\sqrt{\epsilon}}{z} \cdot \frac{1}{\sqrt{x_i}} \quad [m] \tag{B.35}$$

with ϵ relative permittivity, z number of charges of the majority defect species i, and x_i mole fraction of this defect. The specific resistivity of an electron conducting material is about $\varrho[\Omega m] \approx 10^{-8}/n_i$ where n_i means mobile electrons per atom in the semiconducting phase [B.1]. Inserting this in (B.35) yields

$$\rho \approx \frac{10^{-4} \cdot \lambda^2 \, [\text{nm}]}{\epsilon/z^2} \qquad (B.36)$$

or

$$\lambda[\text{nm}] \approx \frac{\sqrt{\epsilon}}{z} \cdot 100\sqrt{\rho\,[\Omega m]}. \qquad (B.37)$$

These estimates demonstrate that even in semiconducting layers with comparatively low specific resistivity in the range of $10^{-2}\Omega m$ the space charge densities become small enough to enable the existence of a Mott equilibrium potential in oxide layers of a few nm thickness at ambient temperature. On the other hand, it can easily be shown that in materials with a specific resistivity of a few $[\Omega m]$ the electronic flux rates can be considerably higher than the maximum oxidation rates detectible by experimental techniques.

A very interesting consequence of the present estimations is the finding that the fast initial branch of metal oxidation, where the reaction rates are strongly enhanced by the Mott field, is not necessarily a process that requires tunneling of electrons through a highly insulating surface layer. Electronic currents in semiconducting substances such as transition metal suboxides are able as well to establish a Mott equilibrium potential across relatively thick oxide films and high oxidation rates should be found also with moderate electronic current densities of about one A/m^2 (Sect. C.3).

B.5 Equilibria in Oxide Layers

In this section equilibria on both interfaces of an oxide layer are considered and their mutual interactions. Total equilibrium is impossible in an oxide layer growing on a metal surface in contact with the gas phase. In Fig. B.7 the equilibria at both interfaces are shown for an oxide layer much thicker than the screening length on both sides. It is assumed that on the metal/oxide interface metal interstitials or oxygen vacancies are formed as discussed in Sect. B 3. The defects shall be donors for electrons and a positive contact potential V_1 exists. The Fermi-level in the oxide is fixed above the donor levels. At the oxide/gas interphase electron acceptor levels are expected to be present in the oxide since excess oxygen atoms as interstitials or metal vacancies are a realistic assumption for the oxide composition in contact with the oxygen of the gas atmosphere. Furthermore, low chemisorption acceptor levels W shall exist on the oxide surface. Also there a positive contact potential shall be present for electrons with respect to the surface states. The Fermilevel E'_F in the surface region of the oxide is fixed below the acceptor levels.

Total equilibrium cannot exist in such a system unless one of the two phases, metal or oxygen, disappears. Then the equilibrium state present on one side of the oxide scale can spread over the whole layer. If the electronic flux is assumed to be very fast compared to the flux of ionic defects, the Fermi-levels on both sides are

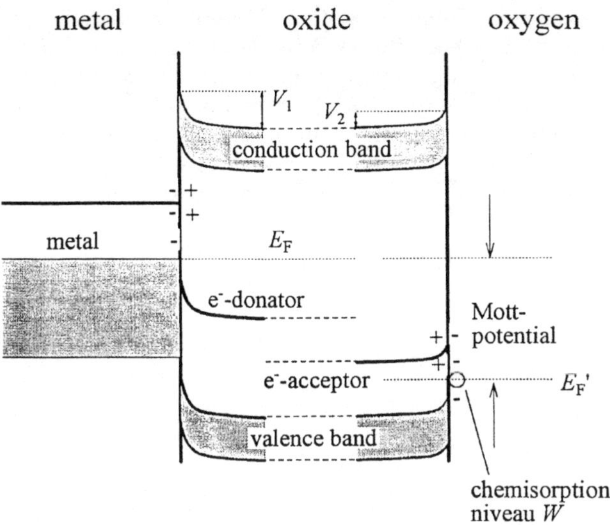

Fig. B.7 Electronic equilibria in an oxide layer on a metal, schematic

equilibrated and across the oxide scale a potential difference V exists which corresponds about to the Mott potential W/e.

Now, what happens really in a growing thick oxide layer? This depends on the relative mobilities and concentrations of the defect species formed on both sides of the layer. The faster flux is more efficient and, therefore, it dominates the reaction mechanism. If the positive defects formed at the metal/oxide interface provide the higher ionic flux the reaction front is the oxide surface and if negative defects formed at the oxide surface are faster new oxide molecules are formed at the metal/oxide interface.

The ionic defects can react with each other in the bulk of the sample and produce new oxide molecules from interstitials or voids from vacancies. An interstitial and a vacancy can also be considered as a Frenkel defect and annihilate. In Wagner's theory such internal equilibria are discussed but for structural reasons they may not be very realistic inside a perfect oxide crystal. Especially in not very thick layers the defects formed at one of the interfaces will migrate directly to the opposite interface and be annihilated there by the oxide formation reaction.

The next problem arises when the oxide layer is thinner than the screening length of the defects. Then the space charges play only a minor role (Sect. B.3) and the potential difference is determined by the Mott potential as long as the electronic currents are fast enough to establish equilibrium. Otherwise, the electric field is established by coupled currents (Appendix D.3). This condition is determined by kinetics and has nothing to do with equilibria. The models developed by Fromhold (Appendix D.1) are derived by using this criterion.

Last but not least one has to keep in mind that a contact potential at the interface of a MOS structure, which is balanced by a space charge in the oxide phase, cannot exist if the coating is much thinner than the screening length (Appendix B.2). But a Mott potential can be established and attain easily values of about one volt.

This short overview on benefits and limitations of equilibrium considerations in reaction kinetics may demonstrate that equilibria between charged particles are important for the understanding of details of the system, but in the reaction mechanism they are frequently overrun by more powerful principles of kinetics. This point of view is different from standard treatments in semiconductor physics, where only electronic transport is considered. In contrast to this situation oxidation kinetics deals with coupled currents and our attention has to be focused on both, fluxes of electrons and of ions.

C Electronic Currents

Electronic currents are very important partial processes in low temperature oxidation models. Their magnitude determines whether the Mott equilibrium potential is established and produces high electric fields inside the oxide layer. This field enhances the ionic fluxes and the oxidation rates are increased drastically. Especially tunneling of electrons from the metal phase to the oxide surface plays a crucial role during the initial stage of low temperature oxidation. But also ohmic currents in semiconductors can attain values high enough to produce a Mott field because of the very small thickness of the oxide layers.

C.1 Tunneling

Quantum mechanical tunneling of a particle through a barrier is described by the relation

$$I = N_{\text{source}} \cdot TP \cdot N_{\text{acceptor}}. \tag{C.1}$$

The flux is proportional to the number of particles in occupied sites in front of the barrier, N_{source}, the tunneling probability of the particle, TP, and the number of empty acceptor sites, N_{acceptor}. The tunneling probability of an electron through a potential barrier of the hight U and the thickness l is given as [C.1]

$$TP = \exp\left(\frac{-2\sqrt{2m_{\text{e}}}}{\hbar} l\sqrt{U}\right) \tag{C.2}$$

with m_{e} effective mass of the electron. In this equation it is assumed that the initial and final energy states are at the same level (resonance tunneling) and the shape of the energy barrier is rectangular.

If electrons from the conduction band of the metal phase tunnel through an oxide scale the situation is different and more complex. In Fig. C.1a the energetic situation is schematically shown when no electric field exists inside the oxide layer. The bottom of the conduction band in the oxide is assumed to be U eV above the Fermilevel of the electrons in the metal phase and the acceptor levels for electrons on the oxide surface shall be W eV below the Fermi-level. The Mott potential W is a negative quantity. In Fig. C.1b electronic equilibrium is established. That means some of the acceptor sites at the surface are occupied by electrons and cause a potential difference $-V$ produced across the oxide scale as a consequence of the Mott potential. The energy of electrons at the surface is in-

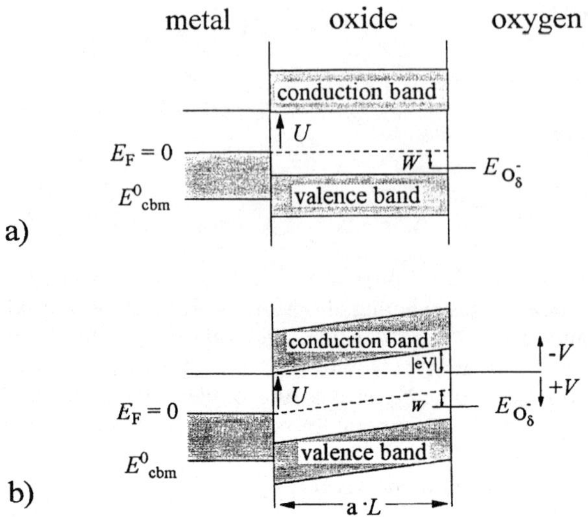

Fig. C.1 Electronic structure of a metal covered with an oxide layer. (a) without field and (b) with field

creased by the amount $|eV| = -eV$. The electric field changes the shape of the potential barrier. The tunnel probability is then given by an expression

$$TP = \exp\left(\frac{-2\sqrt{2m_e}}{\hbar} \int_0^l \sqrt{U - eV(x)}\, dx\right) \tag{C.3}$$

which can be replaced in the limit $V \ll U$ by

$$TP \approx \exp\left(\frac{-2\,l\sqrt{2m_e}}{\hbar} \cdot \sqrt{U - \frac{eV(x)}{2}}\right). \tag{C.4}$$

The density of electrons in the conduction band of the metal, N_{source}, as a function of the distance ΔE from the Fermi level is given mainly by the Fermi function

$$N_{source} = N^0(E)[1 + \exp(\Delta E/kT)]^{-1}. \tag{C.5}$$

The number of acceptor sites, $N_{acceptor}$, can be represented by the coverage of the oxide surface with physisorbed O_2 molecules if O_2^- are assumed to be the only charged chemisorption species.

The equations mentioned so far describe the electron partial flux from an exactly defined electron level inside the metal, for instance lying ΔE eV below the Fermilevel, to an acceptor level on the surface with exactly the same energy. In this limit only electrons from a level $\Delta E = W + eV$ are able to penetrate the oxide skin if the chemisorption level or Mott potential W has a fixed negative value. It has been discussed more detailed elsewhere [C.2–4] that models where the tunneling process is treated within the limit of resonance tunneling can explain most features of the oxidation mechanism caused by tunnel currents through an oxide scale. They demonstrate, for example, that an equilibrium potential V is established with a value close to $-W/e$. At higher negative potential differences V the surface level is above the Fermi-level where the electronic states in the metal conduction band are almost empty. Then the electronic forward flux is decreased drastically whereas the reverse current is increased because of the large number of empty sites now available.

For quantitative calculations resonance tunneling is not a very realistic assumption. The large number of electrons between the resonance level $-(W + eV)$ and the Fermi- level are not able to contribute to the electron flux despite their much higher tunnel probabilities because of the violation of the resonance condition. However, if electron energy levels on the surface are introduced above the chemisorption level W the electrons can be trapped there and emit their excess energy as phonons or by other processes. Under such conditions all electrons in the metal conduction band with energies higher than $-(W + eV)$ can contribute to the tunnel forward current. This conception of possible electron-phonon interactions requires a more complex mathematical treatment [C.3] which finally yields an expression of the form

$$J_{tf} = \alpha_1 \theta_{O_2} kT \cdot \exp[-F_1(U)] \left[F_2(U, T) - \int_{E^0_{cbm}}^{E_{ox}} F_3(E_x, T) dE_x \right]. \qquad (C.6)$$

α_1 is a constant and F_1 to F_3 are functions of the parameters indicated. E_{ox} is written for $(W + eV)$ and E^0_{cbm} is the electron energy at the bottom of the conduction band in the metal. The expression (C.6) is solved numerically in the models of Chap. 5. This more realistic approach guaranties that the tunnel currents are big enough to fit experimental data and the tunnel flux becomes larger when the shape of the energy barrier is changed from a rectangle to a trapezium for $V >< 0$ (Fig. C.1).

The tunnel current in reverse direction is described by a similar expression,

$$J_{tr} = \alpha_2 \theta_{O_2^-} \cdot \exp[-F_4(U, V, E_{ox})] \frac{1}{1 + \exp(-E_{ox}/kT)}. \qquad (C.7)$$

α_2 is again a constant and F_4 a function of energy terms.

The equations for the tunnel fluxes through oxide scales have to meet two requirements in the models:

(i) The equation of the total flux in forward and reverse direction must yield a zero net flux if the Mott equilibrium potential is attained.

(ii) The tunnel currents must decrease exponentially with increasing layer thickness.

The kinetics of the initial stage of the oxidation processes, where field-enhanced ionic transport prevails, cannot be measured quantitatively by experiment. Therefore, a very careful analysis of the consequences arising from the use of different approaches for the description of tunnel currents is of minor practical relevance. Especially considerations on possible effects of the shape of the potential wall on the tunnel current may not contribute to a better understanding of experimental data.

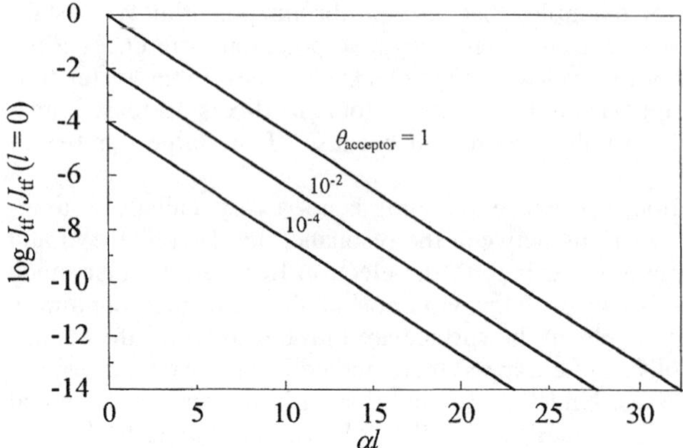

Fig C.2 Tunnel forward current as a function of the oxide layer thickness and the number of acceptor sites on the surface

For more qualitative considerations one has to keep in mind that the tunnel currents are represented generally by equations of the type

$$J_{tf} = \theta_{acceptor} \cdot J(l = 0) \cdot \exp(-\alpha \cdot l). \tag{C.8}$$

The parameter α is an important parameter of the oxidation system. It is increased with increasing values for the energy gap U between the Fermi level in the metal and the bottom of the conduction band of the oxide and the effective mass m_e of the electrons. For large values of the constant α the tunnel flux decreases very fast with increasing layer thickness l (Fig. C.2). At the end of the initial stage of the oxidation process the tunnel current becomes rate controlling. The constant α determines the thickness of the oxide layer formed in the first stage. When the tunnel forward current is no longer able to sustain the Mott equilibrium potential, the fast ionic current breaks down and the layer growth is retarded to low values. The slope of the logarithmic branch in the second stage of the oxidation curves is determined by the constant α (Sect. 5.11).

C.2 Hopping Mechanism

When the thickness of the oxide scale exceeds a critical value, the tunnel currents break down because of their exponential dependence on the layer thickness and another mechanism for the electronic current must become active to guarantee charge neutrality. If the mean free path of electrons in the conduction band of the oxide is assumed to be large compared with the layer thickness l , the thermally activated electronic forward flux can be treated as jumps of electrons over an activation barrier of height $A = U - eV$ and for the reverse flux $A = U + W$ holds (Fig. C.1). The mathematical treatment of the problem is very similar to thermal emission of electrons from metals according to the Richardson equation [C.5]. This relation has to be complemented only by the donor and acceptor sites for electrons on the oxide surface, θ_{acceptor} and θ_{donor}, for example, O_2 and O_2^-. For the forward flux one can write [C.3]

$$J_{\text{thf}} = \theta_{O_2} \frac{m_2}{2\pi^2\hbar^3} (kT)^2 \exp\left(\frac{-(U + (|eV|eV)/2)}{kT}\right) \qquad (C.9)$$

and for the reverse flux

$$J_{\text{thr}} = \theta_{O_2^-} \frac{m_e}{2\pi^2\hbar^3} (kT)^2 \exp\left(\frac{-[U + W(|eV| + eV)/2]}{kT}\right). \qquad (C.10)$$

C.3 Semiconduction

Electronic currents in semiconductors are described by the equation

$$J = n \cdot b \cdot E = n_0\, b \exp\left(\frac{-\Delta U}{kT}\right) \cdot E. \qquad (C.11)$$

with n number of electrons, ΔU distance of the energy gap, and b mobility. If the specific resistivity of a material exceeds the value of one Ωm the material is defined to be semiconducting and under normal conditions only small electronic currents are passing such materials. However, whether a flux is considered large or small depends on the problem of interest. In oxidation models electronic currents must be compared with the growth rate of the oxide. One monolayer of ions is equivalent to about 10^{19} electrons/m^2 or to about 0.3 nm of an oxide scale. Thus, an electronic current density of $5 \cdot 10^{-1}$ A/m^2 corresponds to a reaction rate of one ML/s or to about 30 nm/min, which is a relatively fast process even in the initial stage of low temperature oxidation. It will be demonstrated below that a current density i of one A/m^2 can be produced by a Mott potential W of about one eV (Fig. C.1) across a 10 nm thick layer even if the specific resistivity of the oxide is as high as

$$\rho_{\text{max}} = U/il = 10^8\,\Omega\text{m}. \qquad (C.12)$$

This is in an order of magnitude frequently found with transition metal oxides [C.6, 7]. According to this estimate it can be concluded that the specific electronic conductivity of semiconducting oxide scales is in many oxidation systems high enough to yield growth rates as observed during the initial stages of low temperature oxidation. The temperature dependence of the current of semiconducting materials is described by the same exponential factor as in the hopping model but the electronic current depends here also on the layer thickness l.

In the subsequent semiqantitative discussion the electronic current is simulated by the flux equation for charged species in the homogeneous field limit (D.16) for electrons in the conduction band of the semiconductor. This approach is a simple approximation but it gives an idea how flux equations for electrons in semiconducting oxides in the high field limit may look like. Equilibrium shall be established between electrons in the metal and electrons in the conduction band of the semiconductor at the metal/oxide interface $x = 0$ and between the chemisorption acceptor sites on the oxide surface and conduction band electrons at the position $x = l$.

At the metal/oxide interface the equilibrium condition

e (metal, Fermilevel) = e (cond.)

holds. The equilibrium concentration of electrons in the conduction band can be written as

$$\ln[\theta_{e(cond.)}(0)] = \frac{1}{kT}\left(-\mu^0_{e(cond.)} + \mu^0_{e(metal)} - U\right). \tag{C.13}$$

The standard potential of electrons in the conduction band of the metal can be set zero (A.12) and the standard potential in the conduction band is chosen here so that the activity a becomes $\theta_{e(cond.)}$. According to the definitions presented in Sect. B.4.1 the electronic equilibrium at the oxide surface with O_2^- acceptor sites

e (cond.) + O_2 (phys) $\rightarrow$ O_2^-

is given by the relation

$$\ln[\theta_{e(cond.)}(l)] = \ln\left(\frac{\theta_{O_2^-}}{\theta_{O_2}}\right) + \frac{1}{kT}\left(\mu^0_{O_2^-} - \mu^0_{O_2(phys)} - \mu^0_{e(cond.)} - U\right). \tag{C.14}$$

$\theta_{e(cond.)}(l)$ is the density of electrons in the conduction band of the oxide at the oxide surface.If the electronic current in the conduction band is described by the equation for the homogeneous field limit (D.16) one obtains with $E^0 = -V/l$

$$J_e = 2D^* \sinh\left(\frac{eVa}{2lkT}\right)\left(\frac{\theta_e(l) - \theta_e(0)\exp(eV/kT)}{1 - \exp(eV/kT)}\right) \tag{C.15}$$

with $D^* = \nu \exp(-A/kT)$. Inserting (C.13) and (B.25) in (C.14) yields

$$\theta_{e(cond.)}(l) = \theta_{e(cond.)}(0)\exp\left(\frac{e\Delta\varphi}{kT}\right) = \theta_{e(cond.)}(0)\exp\left(\frac{eV_{eq,Mott}}{kT}\right) \tag{C.16}$$

and (C.15) can be written as

$$J_e = 2D^* \sinh\left(\frac{eVa}{2lkT}\right)\theta_e(0)\left(\frac{\exp\left(\frac{eV_{eq,Mott}}{kT}\right) - \exp\left(\frac{eV}{kT}\right)}{1 - \exp\left(\frac{eV}{kT}\right)}\right). \tag{C.17}$$

In (C.16) the equilibrium potential $\Delta\varphi$ has been replaced by the symbol $V_{eq,Mott}$. Equation (C.17) shows that J_e becomes zero for $V = V_{eq,Mott}$. Both quantities are then negative. In the low field limit instead of (C.17) the much simpler equation

$$J_e = \frac{De}{kT}\theta_e(0)\frac{(|V_{eq,Mott}| - |V|)}{l} \tag{C.18}$$

can be used. If no potential V exists across the oxide layer, an electronic diffusion flux from the metal phase to the oxide surface exists which is originated by the much lower electron density in the conduction band at the surface. The charges piled up at the surface cause a backward electron current, which finally builds up the Mott equilibrium potential if no other charge fluxes exist. The exchange current density in this equilibrium state is given by the term $i = V_{eq,Mott}/\rho l$. If this current is much larger than the ion current then a Mott equilibrium potential can be established by normal electronic currents in semiconducting layers which are thinner than the shielding length λ.

D Ionic Currents

D.1 Basic Equations

A general description of ionic currents is based on hopping models where the migration activation energy A is affected by an applied electric field E [D.1]. A is increased or decreased by an additional quantity $z_i eaE/2$. The distance of the saddle point of the activation barrier from the equilibrium site of the ion is $a/2$ (Fig. D.1). The flux J across the barrier is then given as

$$J_i = J_f - J_r = \theta_f \nu \exp\left(\frac{-(A - z_i eaE/2)}{kT}\right) - \theta_r \nu \exp\left(\frac{-(A + z_i eaE/2)}{kT}\right) \quad \text{(D.1)}$$

with ν hopping frequency, θ_f, θ_r concentration of mobile ions per m^2 in sites for forward and reverse jumps. This flux equation is used for numerical calculations and the basis for approximations valid under simplified conditions. Such limits can be derived for constant concentration where $\theta_f = \theta_r$ holds or if the electrical field is very large or very small.

D.1.1 Zero Concentration Gradient

In the limit $\theta_f = \theta_r$ (D.1) can be written as

$$J_i = 2\theta_i \nu \exp(-A/kT) \cdot \sinh(z_i eEa/2kT). \quad \text{(D.2)}$$

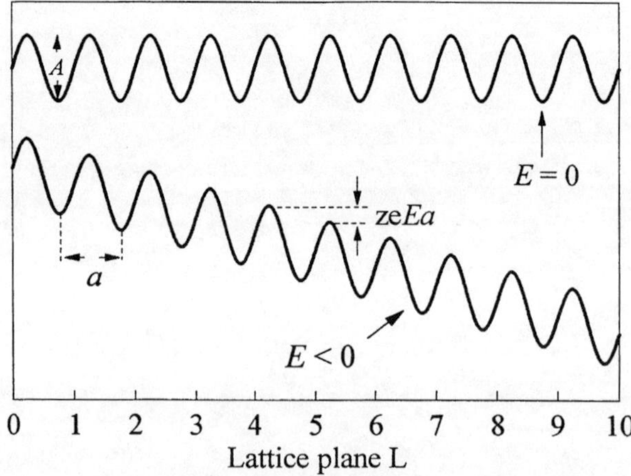

Fig D.1 Potential energy of a charged particle without and with electric field

This expression can be simplified further for very large or very small fields:
high field limit: $|E| \gg 2kT/z_i e\, a$

$$J_i = \theta_i v \exp(-A/kT) \cdot \exp(z_i e E a/2kT);$$ (D.3)

low field limit: $|E| \ll 2kT/z_i e\, a$

$$J_i = \frac{z_i e a}{2kT} \cdot \theta_i \cdot v \exp(-A/kT) \cdot E = \kappa \cdot E.$$ (D.4)

J_i is given here in monolayer equivalents per s. The electric current density in $[A/m^2 s]$ then becomes

$$J_i' = z_i e J_i = \frac{1}{\rho} \cdot E$$ (D.5)

or

$$\frac{1}{\rho} = \frac{z_i^2 e^2 a[m]}{2kT} \cdot \theta_i[m^{-2}] \cdot v \cdot \exp(-A/kT).$$ (D.6)

Equation (D.5) represents Ohms law for electric currents in low electric fields.

D.1.2 Zero Field Current

In the limit $E = 0$ (D.1) becomes

$$J_i = (\theta_f - \theta_r) v \exp(-A/kT).$$ (D.7)

With the definitions $c(x) = \theta/a$, $D = a^2 v \exp(-A/kT)$ and

$$\frac{\partial c(x)}{\partial x} = -\frac{(\theta_f/a - \theta_r/a)}{a} = -\frac{(\theta_f - \theta_r)}{a^2},$$

Ficks first law is obtained

$$J = -D\, \partial c(x)/\partial x.$$ (D.8)

D.1.3 High Field Transport Equation

For the treatment of models based on transport mechanisms in semiconductors the detailed discussion of this subject is given in the books of Fromhold [D.1]. Other papers of the same researchers is extremely helpful and provides very efficient equations for computer simulations. Also, he has shown that (D.1) can be written as a differential equation [D.2]

$$J_i(x) = -2D\frac{\partial c}{\partial x}\cosh\left(\frac{z_i e E(x)a}{2kT}\right) + 2\frac{D}{a}c(x)\sinh\left(\frac{z_i e E(x)a}{2kT}\right)$$ (D.9)

with c in m^{-3} and $D = a^2 v \exp(-A/kT)$. This expression is a first order approximation of a Taylor series and, thus, less accurate than the difference notation of the problem in (D.1).

D.1.4 Steady State Current in a Homogeneous Field

According to Poissons' law the electric field inside a phase is changed by space charges.

$$\frac{\partial E}{\partial x} = \frac{e}{a\epsilon\,\epsilon_0}\sum_i \theta_i z_i.$$
(D.10)

Consequently, the field E in (D.1) depends on the local concentrations θ_i and charges z_i of the particles i. If the electric field inside a semiconducting layer is mainly caused by an external field, the space charges play only a minor role. The field can then be considered constant or homogeneous. This limit is approached if

$$\frac{\Delta E}{E_0} = \frac{-e}{a\epsilon\,\epsilon_0 E_0}\int_0^l \sum_i \theta_i(x)z_i dx \approx -\frac{e}{a\epsilon\,\epsilon_0 E_0}\,l\sum_i \theta_i z_i \to 0$$
(D.11)

holds. The flux equation (D.9) can also be written as

$$J_0 = -D'\frac{\partial c}{\partial x} + b'cE_0$$
(D.12)

with the abbreviations

$$D' = 2D\cosh\left(\frac{z_i e E_0 a}{2kT}\right)$$
(D.13)

and

$$b' = b\frac{2kT}{z_i ea}\sinh\left(\frac{z_i e E_0 a}{2kT}\right).$$

The quantities b and D are correlated by the Einstein relation $ze \cdot D = b \cdot kT$.

If space charge effects are disregarded and the steady state approximation $J(x, t) = \text{const.}$ holds (D.9) can be replaced by the expression

$$\frac{d}{dx}\left[c(x)\exp\left(\frac{-b'E_0 x}{D'}\right)\right] = -\frac{J_0}{D'}\exp\left(\frac{b'E_0 x}{D'}\right)$$
(D.14)

and integration between the limits $x = 0$ and $x = 1$ yields

$$J = b'E_0\left(\frac{c(l) - c(0)\exp(b'E_0 l/D')}{1 - \exp(b'E_0 l/D')}\right).$$
(D.15)

The notation of the problem as differential expressions is an approximative solution. The more exact treatment based on difference equations yields another, even simpler, result [Ref.D.1, p. 250]:

$$J_i[\text{ML/s}] = 2D^* \cdot \sinh\left(\frac{z_i e E_0 a}{2kT}\right)\left(\frac{\theta_i(l) - \theta_i(0)\exp(z_i e E_0 l/kT)}{1 - \exp(z_i e E_0 l/kT)}\right)$$
(D.16)

and

$$\theta_i(x) = \theta_i(0) + [\theta_i(l) - \theta_i(0)] \left(\frac{1 - \exp(z_i e E_0 x/kT)}{1 - \exp(z_i e E_0 l/kT)} \right) \tag{D.17}$$

with θ_i in ML $= c_i/a$ and $D^* = \nu \exp(-A/kT)$ in s^{-1}.

Equation (D.16) is a very useful expression for ionic fluxes in semiconductors. If space charge effects can be disregarded the main features of high field effects and the low field limits of Ohms law and of Ficks first law are represented by this formula. The low homogeneous field limit is approximated by

$$J_0 \approx D^* \left(\frac{\theta_i(l) - \theta_i(0)}{l/a} \right) + \theta_i(0) b^* E_0 \tag{D.18}$$

with $b^* = z_i e a\, D^*/kT$. This is the linear combination of Ficks law and Ohms law. The first term holds for the limit $E = 0$ and the second one for $\theta_i(l) = \theta_i(0)$.

As another notation the relation

$$J = -c(x) \cdot \frac{d}{kT} \cdot \nabla \eta \tag{D.19}$$

can be used for fluxes in systems not far from the equilibrium state, which means in low electric fields and for low concentration gradients. With the expression for the electrochemical potential η of an ionic species

$$\eta = \mu^0 + kT(x) + ze\varphi, \tag{D.20}$$

(D.19) can be written as

$$J = -D \left[\nabla c(x) - c(x) \cdot \frac{ze}{kT} E_0 \right]. \tag{D.21}$$

This relation for the low field limit corresponds to the expression (D.18).

D.2 Space Charge Effects

Space charges in a semiconducting film change the electric fields inside the oxide layer. If this considerably affects the reaction mechanism, numeric methods have to be applied for the calculation of the fluxes under steady state conditions. The hopping model (D.1) can also be used here as a basic equation for the transport mechanism. The concentration profile of the moving ions can be defined on each lattice plane n from the concentration of the preceding plane $n - 1$,

$$\theta_i(n) = \theta_i(n - 1)\exp[-z_i e a E(n)/kT] - (J_i/D^*)\exp[-z_i e a E(n)/2kT]. \tag{D.22}$$

The field E(n) is obtained from Poisson's equation.

$$E(n) - E(n - 1) = -\frac{z_i e \theta_i(n - 1)}{\epsilon \, \epsilon_0}. \tag{D.23}$$

The values for the concentrations at both phase boundaries $n = 1$ and $n = L$ must be compatible with the condition that the electric field disappears in the metal and in the gas phase. In the numeric treatment additional loops are needed with variation of the initial value of $\theta_i(1)$ for a given flux J_i for the determination of the concentration profile and makes the procedure time consuming. Incorporation of space charge effects into models of thin film oxide growth does not normally yield much new information. For most systems the diffusion coefficients D are not exactly known and, thus, are almost free fitting parameters. Therefore, only drastic changes of the reaction mechanism caused by space charges can be detected by comparing model calculations with experimental oxidation curves. If they are less than one or several orders of magnitude the effects on rate and time laws remain hidden behind the scatter of experimental data.

D.3 Coupled Currents

Under steady state conditions the flux of charged species must be zero with respect to the charge transfer:

$$\sum_i z_i J_i = 0. \tag{D.24}$$

Otherwise, unrealistic high electric fields would be established within very short time across an oxide layer. In a system with only one ionic and one electronic mobile species the currents are given in the low homogeneous field limit as

$$J_1 = -D_1\left(\nabla c_1 + \frac{z_1 e c_1}{kT}\nabla\varphi\right) \tag{D.25}$$

and

$$J_2 = -D_2\left(\nabla c_2 + \frac{z_2 e c_2}{kT}\nabla\varphi\right).$$

$\nabla\varphi$ must attain a value which guaranties that $z_1 J_1 = -z_2 J_2$ holds. This condition yields

$$z_1 J_1 = -z_2 J_2 = \frac{z_1 D_1 c_1 \cdot z_2 D_2 c_2}{z_1^2 D_1 c_1 + z_2^2 D_2 c_2}[\nabla(z_2 \ln c_1 - z_1 \ln c_2)]. \tag{D.26}$$

The term $(z_1 D_1 c_1 \cdot z_2 D_2 c_2)/(z_1^2 D_1 c_1 + z_2^2 D_2 c_2)$ can be considered as the diffusion coefficient in ambipolar diffusion which plays a central role in Wagner's oxidation theory. For $D_1 c_1 \ll D_2 c_2$ it is about $(z_1/z_2)D_1 c_1$. This demonstrates the well-known fact that the slower of the two fluxes is rate determining. The faster diffusing species 2 establishes a quasi-equilibrium state which is defined by the relation

$$\nabla\eta_2 = \nabla(kT \ln c_2 + z_2 e\varphi) = 0 \tag{D.27}$$

or

$$\ln c_2(x) = \frac{-z_2 e\varphi(x)}{kT}.$$

The value of the term e/kT is 38.7 [V^{-1}] at 300 K. A concentration difference of one decade built up between two positions in an oxide layer by the faster species produces a diffusion potential difference of only 60 mV. This demonstrates that small potential differences of some ten mV retard the mobility of the faster diffusing particle to a rate close to the flux of the slower particle. The small values of diffusion potentials cannot enhance the rate of the slower particle by more than a factor of about two. This is much less than the promotion by a Mott equilibrium potential in the range of one eV which increases the rate of ionic fluxes by orders of magnitude.

E Units, Material Constants

Table E.1. Special units used in the models

quantity	special units	SI units
areal density	monolayer [ML]	$1\mathrm{ML} \approx 10^{19}/\mathrm{m}^2$
distance	lattice distance [a]	$1\mathrm{a} \approx 2.5 \cdot 10^{-10}\mathrm{m}$
energy	electronvolt [eV]	$1\mathrm{eV} = 1.602 \cdot 10^{-19}\mathrm{J}$
mass	atom mass unit [u]	$1\mathrm{u} = 1.66 \cdot 10^{-27}\mathrm{kg}$

The units a and ML are relative units which have different absolute values for each plane in each phase. The length a denotes the distance between two lattice planes parallel to the surface which offer sites for the species under consideration. The total number of sites N available for the species i on the plane per m^2 defines the unit ML. Examples for these quantities are given in Table E.3

Table E.2. Symbols and special units for physical quantities

quantity	symbol	unit
areal concentration	θ	ML
concentration	$c = \theta/a$	ML/a
current density	J	ML/s
distance	l	a
energy	U, V, A	eV
electric field strength	E	V/a
diffusion constant	$d = a^2 \nu \exp(-A/kT)$	a/s
	$D^* = \nu \exp(-A/kT)$	1/s

Table E.3. Conversion factors for a, ML and R for selected oxides [E1]

	Al_2O_3	TiO_2	V_2O_5	Cr_2O_3	Fe_2O3	NiO	Ta_2O5
N in 10^{28} O atoms/m^3	6.9	6.0	5.7	6.0	6.0	6.0	5.6
$a = N^{-1/3}$ in 10^{-10} m	2.4	2.4	2.6	2.5	2.5	2.5	2.6
ML = $N^{2/3}$ in 10^{19}/m^2							
in O-atoms/m^2	1.68	1.54	1.48	1.54	1.54	1.54	1.46
in O_2 molecules/m^2	0.84	0.77	0.74	0.77	0.77	0.77	0.73
$R = (da/dt)/dJ$							
in a/ML(metal)	1.3	0.9	1.1	1.2	1.2	1.8	0.8
in a/ML(oxygen)	2.1	1.8	1.7	1.8	1.8	1.8	1.7

Table E.4. Dielectric constants of selected oxides [E.1]

	dielectric constant
Al_2O_3	10.5–12
TiO_2	40–50
V_2O_5	14
Cr_2O_3	9.2
FeO	16
CoO	13
NiO	9–12
La_2O_3	20.8

Table E.5. Electron energies in selected oxide systems in eV

metal	work function, metal [E.2]	work function, oxide [E1]	band gap, oxide [E.1]	energy U
Al	4.3	4.8	3.0	2.5
Cr	4.5		4.8	
Cu	4.6	4.7	1.8	1.7
Fe	4.7		2.1	
La	3.5	4.0	5.4	4.9
Mn	4.1		1.3	
Ni	5.1	5.5	1.8	1.4
Ta	4.3	4.6		
Ti	4.3	6.2	3.1	1.2
V	4.3		0.5	
Zn	4.4	4.6 [E.3]	3.2	3.0

Symbols

A_{diff}	activation energy for the diffusion of atoms or defects
$A_{\text{ph,ch}}$	activation energy for the transition from physisorption to chemisorption
$A_{\text{ch,ss}}$	activation energy for the transition from chemisorption to the subsurface
A_{r+}	activation energy in rate constants of the reaction r in forward direction
a_{i}	activity of the species i
a	distance between two planes occupied by defects
b	mobility of a migrating species
c_{i}	concentration of the species i in m^{-3}
D	D_0 exp($-A/kT$): diffusion constant (Table E.2)
e	elementary charge
E_0	electric field strength of a homogenous field in the oxide
$E(L)$	electric field strength between the planes L-1 and L
$E(0) = 0$	electric field strength in the metal
E_{cbm}^0	electron energy at the bottom of the conduction band in the metal
E_F	Fermi energy in the metal
$E_{O_\delta^{\beta-}}$	energy of an electron in the chemisorbed particle $O_\delta^{\beta-}$ referred to the bottom of the valence band of the metal
F_f, F_r	factors in the expression for electronic currents
f_{c}	ratio chemisorption forward flux to maximum diffusion flux
f_{s}	ratio surface penetration forward flux to maximum diffusion flux
G^α	Gibbs free energy of a phase α
g	molal free enthalpy
h	molal enthalpy
$\hbar$	Planck's constant
$J_{\text{ch,ss}}$	partial flux from chemisorption to the bulk subsurface
$J_{\text{ph,ch}}$	partial flux from physisorption to chemisorption
$J_{\text{diff,max}}$	maximum value of the diffusion flux
J_{e}	electronic current
$J_i(L)$	current of the defect i for a layer L lattice planes thick
J_{ion}	ion current
J_{tf}	forward tunnel current
J_{tr}	reverse tunnel current
J_{thf}	forward thermal electron current
J_{thr}	reverse thermal electron current
k	Boltzmann's constant
k^0	pre-exponential factors in rate constants
k_{el}	constant of proportionality for electronic currents
k_{r+}, k_{r-}	forward and reverse rate constants of the reaction r
k_{tf}	constant of proportionality for the forward tunnel current

k_{tr}	constant of proportionality for the reverse tunnel current
k_{thf}	constant of proportionality for the forward thermal current
k_{thr}	constant of proportionality for the reverse thermal current
K_{phys}	equilibrium constant of physisorption
K_r	equilibrium constant of the reaction r
K_S	$= x/p^{1/2}$, Sieverts' constant
l	layer thicknes $l = La$ in m
L	oxide layer thickness in lattice planes
L_{max}	maximum number of oxide layers for a calculation run
m	particle mass
m_e^0	rest mass of the electron
m_e	effective electron mass in an oxide
M^{z^+}	metal interstitial, positive ion
$[M]^{z^-}$	metal vacancy, negative defect
N_i	amount of gas molecules i absorbed in ML
N	number of defect sites per lattice plane in m^{-2}
N_i	number of defects i per lattice plane in m^{-2}
O^{z^-}	oxygen interstitial, negative ion
$[O]^{z^+}$	oxygen vacancy, positive defect
p	pressure
p_i	partial pressure of the species i
r	index for a chemical reaction
r	reaction probability
R	constant of proportionality between the ion current and the oxide growth rate
s	sticking probability
S	surface area of samples
t	time
T	absolute temperature
U	energy gap between the Fermi level in the metal and the bottom of the conduction band in the oxide
v_{r^+}	reaction rate, forward direction
v_{r^-}	reaction rate, backward direction
V	potential difference between the metal and the oxide surface
V	volume of a sample
W	energy gap between the chemisorption level and the metal Fermi level
x	space coordinate
x_i	mol fraction of the component i
x, y	stoichiometric factors in the oxide formula $M_x O_y$
z	$= |z_i|$, number of charges of a defect ($z \geq 0$)
z_i	charge number of the defect i ($z_i \Longleftrightarrow 0$)
α	accommodation coefficient
β	number of electrons in the chemisorbed particle $O_\delta^{\beta-}$
γ_i	activity coefficient of the component i
γ_m	$= m_e/m_e^0$, ratio of effective mass of an electron in an oxide to electron rest mass
δ	number of oxygen atoms in the chemisorbed particle
ΔG	Gibbs free reaction enthalpy

ΔG^0	standard Gibbs free enthalpy of the reaction r
ΔH	reaction enthalpy
ΔH^0	standard reaction enthalpy
Δm	mass gain of a film on a quartz crystal microbalance
ΔS^0	standard reaction entropy
ϵ_0	permittivity of vacuum
ϵ_{ox}	relative permittivity of an oxide
η_i	electrochemical potential of the species i
θ_{chem}	sum of the areal charge densities of chemisorbed particles
$\theta_e(0)$	areal concentration of the compensation charge at the metal/oxide interface $l = 0$
$\theta_i(n)$	$= N_i/N$, areal concentration of the defect i at the plane n
$\theta_{O_\delta^{\beta-}}$	areal density of the chemisorbed particle $O_\delta^{\beta-}$
$\theta_{O_2(phys)}$	areal density of the physisorbed oxygen
μ_i	chemical potential of the species i
μ_i^0	chemical standard potential of the species i
ν	hopping frequency

References

Chapter 2

2.1 S. Gladstone, K. J. Laidler, H. Eyring: *The Theory of Rate Processes* (McGraw-Hill, New York 1941)

2.2 K. J. Laidler: *Theories of Chemical Reaction Rates* (McGraw-Hill, New York 1965)

2.3 G. B. Skinner: *Introduction to Chemical Kinetics* (Academic, New York 1974)

2.4 J. L. Latham, A. E. Burgess: *Elementary Reaction Kinetics* (Butterworths, London 1977)

2.5 E. Fromm, E. Gebhardt (eds): *(Springer, Berlin, Heidelberg 1976)*

2.6 O. Kubaschewski, C. B. Alcock: *Metallurgical Thermochemistry*, 5th edn., (Pergamon, London 1979)

2.7 C. H. P. Lupis: *Chemical Thermodynamics of Materials* (North-Holland, Amsterdam 1983)

2.8 E. T. Turkdogan: *Physical Chemistry of High Temperature Technology* (Academic New York, 1980)

2.9 *JANAF Thermochemical Tables*, 2nd edn. (NSRDS Natl. Bureau of Standards, Washington 1971)

2.10 G.Hörz: Kinetik und Mechanismen, in *Gase und Kohlenstoff in Metallen*, ed. by E. Fromm, E. Gebhardt (Springer, Berlin, Heidelberg 1976) pp. 84–201

2.11 H. J. Kreuzer, Z. W. Gortel: *Physisorption Kinetics*, Springer Ser. Surf. Sci., Vol 1 (Springer, Berlin, Heidelberg 1986)

2.12 K. Kern, G. Cosma: Pysisorbed rare gas adlayers, in *Chemistry and Physics of Solid Surfaces VII*, ed. by R.Vanselow, R. F. Howe, Springer Ser. Surf. Sci., Vol. 10 (Springer, Berlin, Heidelberg 1988) pp. 65–108

2.13 Z. W. Gortel, H. J. Kreuzer: Nonequilibrium desorption of physisorbed atoms, in *Kinetics of Interface Reactions*, ed. by M. Grunze, H. J. Kreuzer, Springer Ser. Surf. Sci., Vol 8 (Springer, Berlin, Heidelberg 1987) pp. 44–53

2.14 M. C. Desjonquères, D.Spanjaard: *Concepts in Surface Science*, 2nd edn. (Springer, Berlin, Heidelberg 1997) pp. 381–499

2.15 G. Ehrlich: Activated chemisorption, in *Chemistry and Physics of Solid Surfaces VII*, ed. by R. Vanselow, R. F. Howe, Springer Ser. Surf. Sci., Vol 10 (Springer, Berlin, Heidelberg 1988) pp. 1–64

2.16 M. Grunze, H. J. Kreuzer (eds.): *Kinetics of Interface Reactions*, Springer Ser. Surf. Sci., Vol 8 (Springer, Berlin, Heidelberg 1987)

2.17 G. Ertl: Reactivity of surfaces, in *Cemistry and Physics of Solid Surfaces VIII*, ed, by R. Vanselow, R. Howe, Springer Ser. Surf. Sci., Vol 22 (Springer, Berlin, Heidelberg 1990) pp. 1–22

2.18 H. Lüth: *Surfaces and Interfaces of Solid Materials*, 3rd edn. (Springer, Berlin, Heidelberg 1995)

2.19 R. J. Madix (ed): *Surface Reactions*, Springer Ser. Surf. Sci., Vol 34 (Springer, Berlin, Heidelberg 1994)

2.20 M. Lannoo, P. Friedel: *Atomic and Electronic Structure of Surfaces, Theoretical Foundation*, Springer Ser. Surf. Sci., Vol 16 (Springer, Berlin, Heidelberg 1991)

2.21 V. F. Kislev, O. V. Krylov: *Electronic Phenomena in Adsorption and Catalysis on Semiconductors and Dielectrics*, Springer Ser. Surf. Sci., Vol 7 (Springer, Berlin,Heidelberg 1987)

2.22 V. F. Kislev, O. V. Krylov: *Adsorption and Catalysis on Transition Metals and their Oxides*, Springer Ser. Surf. Sci., Vol 9 (Springer, Berlin, Heidelberg 1989)

2.23 H. J. Freund, E. Umbach (eds.): *Adsorption on Ordered Surfaces of Ionic Solids and Thin Films*, Springer Ser. Surf. Sci., Vol 33 (Springer, Berlin, Heidelberg 1991) pp. 103–220

2.24 W. Göpel, T. A. Jones, M. Kleitz, J. Lundström, T. Seiyama (eds): Chemical and biochemical sensors, Pt. 1, in *Sensors, a Comprehensive Survey*, Vol. 2, ed. by W. Göpel, J. Hesse, J. N. Zemel (VCH, Weinheim, 1991)

2.25 H. J. Kreuzer: Theory of desorption kinetics, in *Chemistry and Physics of Solid Surfaces VII*, ed. by R. Vanselow, R. F. Howe, Springer Ser. Surf. Sci., Vol 10 (Springer, Berlin, Heidelberg 1988) pp. 259–282

2.26 J. Cranck: *The Mathematics of Diffusion* (Clarendon, Oxford 1990)
2.27 R. Kirchheim: Acta Metall. **29**, 835–844 (1981)
2.28 R. Kirchheim: Acta Metall. **29**, 845–853 (1981)
2.29 R. Kirchheim, F. Sommer, G. Schluckebier: Acta Metall. **30**, 1059–1068 (1982)
2.30 R. Kirchheim: Acta Metall. **30**, 1069–1078 (1982)

Chapter 3

3.1 H. Lüth: *Surfaces and Interfaces of Solid Materials* 3rd edn. (Springer, Berlin, Heidelberg 1995)
3.2 S. Wagener: Brit. J. Appl. Phys. **1**, 255 (1950)
3.3 E. Fromm, H.G. Wulz: J. Less-common Met. **101**, 469–479, (1984)
3.4 H. H. Uchida, H. G. Wulz, E. Fromm: J. Less-common Met. **172–174**, 1076–1083, (1991)
3.5 E. Fromm, H. H. Uchida: J. Less-common Met. **131**, 1–12, (1987)
3.6 G. Z. Sauerbrey: Zeitschrift für Physik **155**, 206 (1959)
3.7 C. Lu, A. W. Czanderna (eds): *Application of Piezoelectric Quartz Crystal Microbalances* (Elsevier, Amsterdam 1982)
3.8 M. Martin, E. Fromm: Thin Solid Films **236**, 199–203 (1993)
3.9 R. M. A. Azzam, N. M. Bashara: *Ellipsometry and Polarized Light* (North Holland, Amsterdam, 1977)
3.10 J. A. Woollam, P. G. Snyder: Variable angle spectroscopic ellipsometry, in *Encyclopedia of Materials Characterization*, ed. by C. R. Brundle, C. A. Evans Jr., S. Wilson (Manning Publ., Greenwich, CT 1992)
3.11 J.A. Woollam, J. Hale, H.W. Yao: In-situ and ex-situ applications of spectroscopic ellipsometry, in *Diagnostic Techniques for Semiconductor Materials Processing*, ed. by O. J. Glembocki, MRS Proc. **324**, 15 (MRS Pittsbourgh, PA 1994)
3.12 E. A. Irene, J. A. Woollam: *In-situ Ellipsometry in Microelectronics*, Materials Research Bulletin **20**, 24 (1995)
3.13 J. N. Hilfiker, D. W. Glenn, S. Heckens, J. A. Woollam: J. Appl. Phys. **79**, 6193 (1996)
3.14 P. G. Snyder, A. Massengale, K. Memarzadeh, J. A. Woollam, D. C. Ingram, P. P. Pronko: Study of ion implanted copper laser mirrors by spectroscopic ellipsometry, in *Beam-Solid Interactions and Transient Processes*, ed. by M. O. Thomson, S. T. Picraux, J. S. Williams MRS Proc. **74**, 535 (MRS, Pittsbourgh, PA 1987)
3.15 C. M. Herzinger, P. G. Snyder, F. G. Celii, Y. C. Kao, D. Chow, B. Johs, J. A. Woollam: J. Appl. Phys. **79**, 2663 (1996)
3.16 A. C. Boccara, C. Pickering, J. Rivory (eds.): *Spectroscopic Ellipsometry*, Proc. First Intern. Conf. on Spectroscopic Ellipsometry (Elsevier, Amsterdam 1993)
3.17 J. W. Mayer, E. Rimini (eds.): *Ion Beam Handbook for Material Analysis* (Academic, New York 1977)
3.18 J. R. Tesmer, M. Nastasi (eds.): *Handbook of Modern Ion Beam Materials Analysis* (Materials Research Society, Pittsburgh, PA 1995)
3.19 B. K. Patnaik, C. V. Barros Leite, G. B. Baptista; E. A. Schweikert, D. L. Cocke, L. Quinones, N. Magnussen: Nucl. Instr. Methods **B35**, 159 (1988)
3.20 W. K. Chu, J. W. Mayer, M. A. Nicolet: *Backscattering Spectrometry*, (Academic, New York 1978)
3.21 Th. Enders, M. Rilli, H. D. Carstanjen: Nucl. Instr. Methods **B64**, 817 (1992)
3.22 D. D. Cohen, E. K. Rose: Nucl. Instr. Methods **B66**, 158 (1992)
3.23 G. Dollinger, T. Faestermann, P. Maier-Komor: Nucl. Instr. Methods **B64**, 422 (1992)
3.24 R. Plieninger, H. D. Carstanjen: Nucl. Instr. Methods, **B125**, 128–132 (1997)
3.25 J. Jamecsny, D. Plachke, H. Carstanjen: In Proc. Int'l Conf. On Microscopy of Oxidation III, ed. by s. B. Newcomb, J. A. Little (Inst. Materials, London, UK 1997) pp. 369–381
3.26 O. Kruse, H. D. Carstanjen: Nucl. Instr. Methods **B89**, 191 (1994)
3.27 H. D. Carstanjen, W. Decker, J. Diehl, Th. Enders, R.M. Emrick, A. Föhl, E. Friedland, D. Plachke, H. Stoll: Nucl. Instr. Methods **B51**, 152 (1990)
3.28 Th. Osipowicz, K. P. Lieb, S. Brüssmann: Nucl. Instr. Methods **B18**, 232 (1987)
3.29 T. P. Russell: Mat. Sci. Rep. **5**, 171 (1990)
3.30 G. P. Felcher, H. You (eds.): Proc. 4th Intern. Conf. on Surface X-ray and Neutron Scattering SXNS-4. Physica **B 221** (1996)
3.31 F. Stangleier, B. Lengeler, W. Weber, H. Göbel, M. Schuster: Acta Cryst. A **48**, 626 (1992)
3.32 A. Stierle, T. Mühge, H. Zabel: J. Mater. Res. **9**, 884 (1994)
3.33 A. Stierle, P. Bödeker, H. Zabel: Surf. Sci. **327**, 9 (1995)

3.34 A. Plech, U. Klemradt, T.H. Metzger, J. Peisl: J. Phys: Condens. Matter **10**, 971–982 (1998)
3.35 D. J. O'Connor, B. A. Sexton, R. St. C. Smart (eds): *Surface Analysis Methods in Materials Science*, Springer Ser. Surf. Sci., Vol 23 (Springer, Berlin, Heidelberg 1992)
3.36 G. A. Somerjai: *Introduction to Surface Chemistry and Catalysis*, (Wiley, New York 1994)
3.37 M. A. Van Hove, W. H. Weinberg, C. M. Chan : *Low-Energy Electron Diffraction*, Springer Ser. Surf. Sci., Vol 6 (Springer, Berlin, Heidelberg 1986)
3.38 H. J. Güntherodt, R. Wiesendanger (eds.): *Scanning Tunneling Microscopy I*, 2nd edn., Springer Ser. Surf. Sci., Vol 20 (Springer, Berlin, Heidelberg 1994)
3.39 R. Wiesendanger, H. J. Güntherodt (eds.): *Scanning Tunneling Microscopy II*, 2nd edn., Springer Ser. Surf. Sci., Vol 28 (Springer, Berlin, Heidelberg 1995)
3.40 C. Bai: *Scanning Tunneling Microscopy and ist Application*, Springer Ser. Surf. Sci., Vol 32 (Springer, Berlin, Heidelberg 1995)
3.41 S. Hüfner: *Photoelectron Spectroscopy*, 2nd edn., Springer Ser. Solid-State Sci., Vol. 82 (Springer, Berlin, Heidelberg 1996)
3.42 L. Reimer: *Scanning Electron Microscopy*, 2nd edn., Springer Ser. Opt. Sci., Vol. 45 (Springer, Berlin, Heidelberg 1998)
3.43 L. Reimer: *Transmission Electron Microscopy*, 4th edn., Springer Ser. Opt. Sci., Vol. 36 (Springer, Berlin, Heidelberg 1997)

Chapter 4

4.1 W. M. Mueller, J. P. Blackledge, G. G. Libowitz: *Metal Hydrides*, (Academic, New York 1968)
4.2 G. Alefeld, J. Völkl (eds.): *Hydrogen in Metals I*, Topics Appl. Phys., Vol 28 (Springer, Berlin, Heidelberg 1978)
4.3 G. Alefeld, J. Völkl (eds.): *Hydrogen in Metals II*, Topics Appl. Phys., Vol 29 (Springer, Berlin, Heidelberg 1979)
4.4 L. Schlapbach: *Hydrogen in Intermetallic Compounds I*, Topics Appl. Phys., Vol 63 (Springer, Berlin, Heidelberg 1988)
4.5 L. Schlapbach: *Hydrogen in Intermetallic Compounds II*, Topics Appl. Phys., Vol 68 (Springer, Berlin, Heidelberg 1992)
4.6 Y. Fukai : *The Metal-Hydrogen System, Basic Bulk Propperties*, Springer Ser. Mater. Sci., Vol. 21 (Springer, Berlin, Heidelberg 1993)
4.7 E. Fromm, G. Hörz: *Int'l Metall. Rev.* **25**, *269 (1980)*
4.8 E. Fromm, E. Gebhardt (eds.): *Gase und Kohlenstoff in Metallen*, (Springer, Berlin, Heidelberg 1976)
4.9 D. E. J. Talbot: *Int'l Metall. Rev.* **20**, *166–184 (1975)*
4.10 H. K. Birnbaum: Mechanisms of hydrogen related fracture of materials, in *Hydrogen Effects on Material Behaviour*, ed. by N. R. Moody, A. W. Thomson, Proc. TMS Meeting, Moran, WY, 1989 (The Minerals Metals & Materials Soc., Warrendale, PA 1990)
4.11 G. Sandrock, S. Suda, L. Schlapbach: Applications, in [Ref 4.5 pp. 197–258]
4.12 H. Uchida, E. Fromm: Z. Metallkde **71**, 85–89 (1980)
4.13 J. A. Rodrigues, E. Fromm: Metallkde **76**, 320–325 (1985)
4.14 E. Fromm, H.G. Wulz: J. Less-Common Met. **101**, 469–479 (1984)
4.15 H. H. Uchida, H. G. Wulz, E. Fromm: J. Less-Common Met. **172-174**, 1076–1083 (1991)
4.16 E. Fromm, H. H. Uchida: J. Less-Common Met. **131**, 1–12 (1987)
4.17 H. G. Wulz, E. Fromm: J. Less-Common Met. **118**, 315–326 (1986)
4.18 H. G. Wulz, E. Fromm: J. Less-Common Met. **118**, 293–301 (1986)
4.19 H. H. Uchida, E. Fromm: J. Less-Common Met. **131**, 125–132 (1987)
4.20 N. Hosoda, H. H. Uchida, E. Fromm: J. Less-Common Met. **172-174**, 824–831 (1991)
4.21 E. Fromm, H. Uchida : J. Less-Common Met. **66**, 77–88 (1979)
4.22 F. Schweppe, M. Martin, E. Fromm: J. Alloys and Compounds **253–254**, 511–514 (1997)
4.23 F. Schweppe: Einfluß von Gasverunreinigungen auf die Kinetik der Wasserstoffaufnahme von Hydridspeicherwerkstoffen; Dissertation, University of Stuttgart (1995); and Fortschrittberichte VDI, Reihe 5, No. 421 (VDI Verlag, Düsseldorf 1996)
4.24 M. Martin, C. Gommel, C. Borkhart, E. Fromm: J. Alloys and Compounds **238**, 193–201 (1996)
4.25 J. Cranck: *The Mathematics of Diffusion*, (Clarendon, Oxford 1990)
4.26 O. Levenspiel: *Chemical Reaction Engineering*, (Wiley, New York 1972) pp. 357–377
4.27 F. Schweppe, M. Martin, E.Fromm: J. Alloys and Compounds **261**, 254–258 (1997)
4.28 J. R. Lacher: Pro. Roy. Soc. (London) Ser. A **161**, 525 (1937)

Chapter 5

5.1 G. Tammann: Z. Anorganische Chemie **111**, 78 (1920)
5.2 N. B. Pilling, R. E. Bedworth: J. Inst. Metals **29**, 529 (1923)
5.3 C. Wagner: Z. physik. Chem. **B21**, 25 (1933)
5.4 C. Wagner: Z. physik. Chem. **B32**, 447 (1936)
5.5 K. Hauffe: *Oxidation of Metals* (Plenum, New York 1965)
5.6 N. Cabrera, N. F. Mott: Rep. Prog. Phys. **12**, 163 (1949)
5.7 A. T. Fromhold Jr: *Theory of Metal Oxidation*, Vol. 1, *Fundamentals* (North-Holland, Amsterdam 1976); Vol. II, *Space Charge* (North-Holland, Amsterdam 1980)
5.8 E. Fromm: Model Calculations of Metal Oxidation at Ambient Temperatures in J. Nowotny, W. Weppner (eds.): *Non-stoichometric Compounds, Surfaces, Grain Boundaries and Structural Defects*, (Kluwer, Drodrecht, Boston 1989) pp. 523–534
5.9 V. Grajewski, E. Fromm: Low-temperature oxidation of metals, in *Interface Segregation and Related Processes*, ed. by J. Nowotny (Trans. Tech. Publ., Zürich 1991) pp. 337–399
5.10 M. Martin, E. Fromm: J. Allogs and Compounds **258**, 7–16 (1997)
5.11 M. Martin: Experimente mit der Schwingquarzwaage zur Oxidation von Eisen-, Aluminium- und Titanfilmen zwischen 50 und 200 °C und Modellrechnungen zur Raumtemperaturoxidation, Dissertation, University of Tübingen (1992); and Fortschrittberichte VDI, Serie 5: Grund- und Werkstoffe Nr. 311 (VDI Verlag, Düsseldorf 1993)
5.12 F. P. Fehlner: *Low-Temperature Oxidation, The Role of Vitreous Oxides* (Wiley, New York 1986)
5.13 H. Chichy, E. Fromm: Thin Solid Films **195**, 147–158 (1991)
5.14 M. Martin, W. Mader, E. Fromm: Thin Solid Films **250**, 61–66 (1994)
5.15 E. Fromm, O. Mayer: Surface Sci. **74**, 259–275 (1978)
5.16 U. R. Evans: *The Corrosion and Oxidation of Metals:Scientific Principles and Practical Applications*, (St. Martins's Press, New York 1960)
5.17 N. F. Mott: Trans.Faraday Soc. **35**, 1175 (1939); **36**, 472 (1940)
5.18 W. Schottky: Zur Frage der rationalen Störstellenbezeichnung, in *Halbleiterprobleme*, **4**, 235 (Vieweg, Braunschweig 1958)
5.19 H. Rickert : *Electrochemistry of Solids* (Springer, Berlin, Heidelberg 1982)
5.20 J. D. Jackson: *Classical Electrodynamics* (Wiley, New York 1975)
5.21 S. Gladstone, K. J. Laidler, H. Eyring: *The Theory of Rate Processes* (Mc Graw-Hill, New York 1941)
5.22 Ch. Kittel: *Introduction to Solid State Physics* (Wiley, New York 1976) p. 244
5.23 V. Grajewski: Modellrechnungen zur Raumtemperatur-oxidation von Metallen und Experimente mit Nickel- und Titanfilmen, Dissertation, University of Stuttgart (1989)
5.24 S. R. Pollack, C. E. Morris: J. Appl. Phys. **35**, 1503 (1964)
5.25 J. E. Boggio, R. C. Plumb: Chem. Phys. **44**, 1081 (1966)
5.26 J. E. Boggio: J. Chem. Phys. **53**, 3544 (1970)
5.27 M. Ronay, E. E. Latta: Phys. Rev. B **32**, 537 (1985)
5.28 N. Tsuda, K. Nasu, A. Yanase, K. Siratori: *Electronic Conduction in Oxides*, Springer Ser. Solid-State Sci., Vol. 94 (Springer, Berlin, Heidelberg 1991)
5.29 P. A. Cox: *Transition Metal Oxides* (Clarendon, Oxford 1992)

Chapter 6

6.1 E. Fromm, H. Uchida: J. Less-Common Metals **66**, 77–88 (1979)
6.2 H. Uchida, E. Fromm: In *Proc. 2nd Int'l Symp. Hydrogen in Metals*, Minakami, Jpn. 1979, ed. by T. Suzuki Transactions Jpn. Inst. Metals **21**, 289–292 (1980)
6.3 H. Uchida, E. Fromm: J. Less-Common Met. **95**, 139–146, 147–152, 153–155 (1983)
6.4 E. Fromm, H. G. Wulz: J. Less-Common Met. **101**, 469–479 (1984)
6.5 H. G. Wulz, E. Fromm: J. Less-Common Met. **118**, 315–326 (1986)
6.6 H. G. Wulz, E. Fromm: J. Less-Common Met. **118**, 293–301 (1986)
6.7 E. Fromm: Z. Phys. Chem. Neue Folge **147**, 61–75 (1986)
6.8 H. G. Wulz, H. Chichy, E. Fromm: J. Less-Common Met. **118**, 303–313 (1986)
6.9 H. H. Uchida, E. Fromm: J. Less-Common Met. **131**, 125–132 (1987)
6.10 E. Fromm, H. H. Uchida: J. Less-Common Met. **131**, 1–12 (1987)
6.11 H. H. Uchida, E. Fromm: Z. Phys. Chem. Neue Folge **164**, 1123–1128 (1989)
6.12 N. Hosoda, H. H. Uchida, E. Fromm: J. Less-Common Met. **172–174**, 824–831 (1991)
6.13 H. H. Uchida, H. G. Wulz, E. Fromm: J. Less-Common Met. **172–174**, 1076–1083 (1991)

6.14 J. A. Rodrigues, E. Fromm: Z. Metallkde **76**, 320 (1985)

6.15 K. Nakamura, H. Uchida, E. Fromm: J. Less-Common Met. **80**, 19–30 (1981)

6.16 L. Schlapbach: *Hydrogen in Intermetallic Compounds II*, Topics Appl. Phys., Vol 68 (Springer, Berlin, Heidelberg 1992)

6.17 G. D. Sandrock, P. D. Goodell, J. Less-Common Met. **104**, 159–173 (1984)

6.18 F. R. Block, H. J. Bahs: J. Less-Common Met. **89**, 77–84 (1983)

6.19 F. Schweppe: Einfluß von Gasverunreinigungen auf die Kinetik der Wasserstoffaufnahme von Hydridspeicherwerkstoffen; Dissertation, University of Stuttgart (1995); and Fortschrittsberichte VDI, Reihe **5**: *No 421 (VDI Verlag Düsseldorf 1996)*

6.20 F. Schweppe, M. Martin, E. Fromm: J. Alloys and Compounds **261**, 254–258 (1997)

6.21 E. Fromm: In *Proc. 2nd Int'l Symp. Hydrogen in Metals*, Minakami, Jpn. 1979, ed. by T. Suzuki Transactions Jpn. Inst. Metals **21**, 285–288 (1980)

6.22 K. Nakamura: Z. Phys. Chem. Neue Folge **116**, 163 (1979)

6.23 E. Fromm, H. Uchida, B. Chelluri: Bcr. Bunsenges. Phys. Chem. **87**, 410–418 (1983)

6.24 J. K. Norskov, F. Besenbacher: J. Less-Common Metals **130**, 475–490 (1987)

Appendices

A.1 G. N. Lewis: *Thermodynamics*, 2nd edn., ed. by K. S. Pitzer, L. Brewer (Mc Graw-Hill, New York 1961)

A.2 O. Kubaschewski, C. B. Alcock, P. J. Spencer: *Materials Thermochemistry*, 6th edn. (Pergamon, Oxford 1993)

A.3 C. Wagner: *Thermodynamics of Alloys*, (Adison-Wesley, Reading, MA 1952)

A.4 I. Prigodine, R. Defay: *Chemical Themodynamics* (Longmans, Green & Co, London 1954)

A.5 E. A. Guggenheim: *Thermodynamics*, 2nd edn. (North-Holland, Amsterdam 1957)

A.6 E. Fromm: Thermodynamik, in *Gase und Kohlenstoff in Metallen* ed. by E. Fromm, E. Gebhardt (Springer, Berlin, Heidelberg 1976)

A.7 F. A. Kröger: *The Chemistry of Imperfect Crystals* (North-Holland, Amsterdam 1964)

A.8 C. H. P. Lupis: *Chemical Thermodynamics of Materials* (North-Holland, New York, Amsterdam 1983)

A.9 J. M. Honig: *Thermodynamics* (Elsevier, Amsterdam 1982)

A.10 H. Rickert: *Electrochemistry of Solids* (Springer, Berlin, Heidelberg 1982)

B.1 C. Kittel: *Introduction to Solid State Physics*, 4th edn. (Wiley, New York 1968) p. 306

C.1 Cunli Bai: *Scanning Tunneling Microscopy and Ist Application* Springer Ser. Surf. Sci.,Vol.32 (Springer, Berlin, Heidelberg 1992)

C.2 V. Grajewski: *Modellrechnungen zur Raumtemperatur-Oxidation von Metallen und Experimente mit Nickel- und Titanfilmen*, Dissertation, University of Stuttgart (1989)

C.3 M. Martin: Experimente mit der Schwingquarzwaage zur Oxidation von Eisen-, Aluminium- und Titanfilmen zwischen 50 und 200 °C und Modellrechnungen zur Raumtemperaturoxidation, Dissertation, University of Tübingen (1992); Fortschrittberichte VDI, Serie 5: Grund- und Werkstoffe Nr. 311, (VDI Verlag, Düsseldorf 1993)

C.4 A. T. Fromhold: *Theory of Metal Oxidation* (North-Holland, Amsterdam 1976) p. 289

C.5 M. C. Desjonquères, D. Spanjaard: *Concepts in Surface Physics* 2nd edn. (Springer, Berlin, Heidelberg 1997) p. 377

C.6 G. V. Samsonov (ed.): *The Oxide Handbook* (Plenum, New York 1973)

C.7 H. H. Kung: Transition metal oxides: surface chemistry and catalysis, in *Studies in Surface Science and Catalysis*, Vol. 45 (Elsevier, Amsterdam 1989)

D.1 A. T. Fromhold: *Theory of Metal Oxidation*, Vol. I, *Fundamentals* (North-Holland, Amsterdam 1976)

D.2 A. T. Fromhold, E. L. Cook: J. Appl. Phys. **38**, 1546 (1967)

E.1 G. V. Samsonov (ed.): *The Oxide Handbook* (Plenum, New York 1973)

E.2 H. B. Michaelson, J. Appl. Phys **48**, 4729 (1977)

E.3 H. Moorman, D. Kohl, G. Heiland: Surf. Sci. **80**, 261 (1997)

Springer and the environment

At Springer we firmly believe that an international science publisher has a special obligation to the environment, and our corporate policies consistently reflect this conviction.

We also expect our business partners – paper mills, printers, packaging manufacturers, etc. – to commit themselves to using materials and production processes that do not harm the environment. The paper in this book is made from low- or no-chlorine pulp and is acid free, in conformance with international standards for paper permanency.

Springer

Printing: Saladruck, Berlin
Binding: Buchbinderei Lüderitz & Bauer, Berlin